SE Structural Engineering
Buildings Practice Exam

Fifth Edition

Joseph S. Schuster, PE, SE

PPI2PASS.COM
A **KAPLAN** COMPANY

Register Your Book at ppi2pass.com

- Receive the latest exam news.
- Obtain exclusive exam tips and strategies.
- Receive special discounts.

Report Errors for This Book

PPI is grateful to every reader who notifies us of a possible error. Your feedback allows us to improve the quality and accuracy of our products. Report errata at **ppi2pass.com**.

SE STRUCTURAL ENGINEERING BUILDINGS PRACTICE EXAM
Fifth Edition

Current release of this edition: 2

Release History

date	edition number	revision number	update
Nov 2017	4	1	New edition. Code updates. Copyright update.
Aug 2018	5	1	New edition.
Jun 2019	5	2	Minor cover updates.

PPI
1250 Fifth Avenue, Belmont, CA 94002
(650) 593-9119
ppi2pass.com

ISBN: 978-1-59126-627-3

Table of Contents

About the Author

Joseph S. Schuster, PE, SE, is a practicing structural engineer licensed in New York, New Jersey, Connecticut, and Illinois. He holds a bachelor of science degree in civil engineering from Cornell University and a master of science degree in structural engineering from Stanford University. Currently, Mr. Schuster works in the Newark office of Thornton Tomasetti, a leading engineering firm. His work is focused on the forensic investigation of failed structural systems, and the repair and adaptive reuse of buildings. Mr. Schuster is a co-author of AISC *Design Guide 15: Rehabilitation and Retrofit*, a reference for the evaluation and strengthening of historic steel construction.

Thornton Tomasetti is an engineering firm that optimizes the design and performance of structures, materials and systems for projects of every size and level of complexity. An employee-owned organization of engineers, scientists, architects and other professionals collaborating from offices worldwide, the firm supports clients by drawing on the diverse expertise of its integrated practices. Thornton Tomasetti is committed to being a sustainable and enduring organization and the global driver of innovation in the architecture, engineering and construction industry.

Preface and Acknowledgments

I wrote *SE Structural Engineering Buildings Practice Exam* to be an accurate representation of the 16-hour Structural Engineering exam (also commonly called the Structural PE exam or the SE exam), which is administered by the National Council of Examiners for Engineering and Surveying (NCEES). Though you will not find any exact exam problems in this book, I have closely followed the NCEES exam specifications so that this practice exam is representative of the SE exam in its content, format, level of difficulty, and length.

It is my hope that all of the problems in this book will be beneficial to your SE exam preparation. The problems may also be useful to anyone preparing for the California Civil Seismic Principles exam or the Civil Breadth and Structural Depth PE exam or to anyone wishing to further his or her understanding of structural engineering concepts.

In this fifth edition, problems have been updated or replaced to ensure that they are up-to-date with the NCEES exam-adopted codes and specifications.

I would also like to thank Nicholas Cramsey, who reviewed all the problems and solutions and provided valuable suggestions for improving the content. His extensive knowledge of the codes and reference standards was a vital resource for me while creating this book. Thanks to Hannah Wilson, who reviewed and edited questions for the latest round of code updates.

Thank you to William J. Kirkham, SE, who reviewed the first edition of this book for technical accuracy and exam appropriateness. His suggestions and comments have much improved this book.

Thank you to Arthur Richard Chianella, PE, who reviewed new content in the third edition for technical accuracy. And to Dr. Ralph Arcena, PE, for reviewing the fourth edition.

Thanks also go to Lindsay Driscoll, who was instrumental in collecting study tips and advice from past examinees.

At PPI, the task of making the vision of a new edition into a reality fell to a product development team that consisted of Nancy Peterson, project manager; Kim Wimpsett, freelance proof reader; Tom Bergstrom, typesetter, technical illustrator, cover designer; Jenny King, editor-in-chief; Grace Wong, director of publishing services; Cathy Schrott, production services manager; Leata Holloway, senior acquisitions editor; and Steve Buehler, director of acquisitions.

While the problems and solutions in this book have been carefully written and reviewed, if you do find an error or have a suggestion, please let me know about it by using the error reporting form on PPI's website at **ppi2pass.com/errata**. Valid submitted errors will be posted to the errata page and incorporated into future printings of this book.

Finally, this book is dedicated to my wife, Heather. Her support and encouragement made this book a reality.

Joseph S. Schuster, PE, SE

Introduction

ABOUT THIS BOOK

This book contains a complete 16-hour practice exam for the Structural Engineering (SE) exam in buildings, including both vertical and lateral exam components. A morning breadth module, covering buildings and bridges topics, is included for each component. This book also contains an afternoon depth module covering buildings topics only. Topics covered reflect the exam specifications identified by the National Council of Examiners for Engineering and Surveying (NCEES) for each component. (See Table 1.)

For the morning breadth modules, problems are written in the same multiple-choice format and level of difficulty as on the SE exam. Problems take an average of six minutes to solve. Breadth problems are scored based solely on the option selected (rather than on the documented solution method, as depth problems are scored). The problems are written to avoid ambiguity and don't require you to make any simplifying assumptions. Problems may be either qualitative or quantitative. Quantitative problems present an introductory statement, followed by a list of required design codes and additional design data. Illustrations showing structural details, dimensions, and loading information are also included. This book provides blank answer sheets for you to record your answers (as you would on the actual exam). A step-by-step solution is included for each breadth problem, so that you can review your work and correct any errors.

For the afternoon depth modules, problems are similar to the morning breadth quantitative problems in that they present an introductory statement, followed by a list of required design codes and additional design data. However, the depth problems aren't multiple-choice; instead, they require you to write essay responses and show your calculations. Like the depth problems on the exam, the depth problems in this book are less straightforward than the multiple-choice problems, sometimes require you to make assumptions, and always require you to show all of your work to receive full credit. (The components of the depth solutions that you would be required to show to receive full credit are presented in blue in this book, and supporting text is black.) You have an average of one hour to solve each of the exam's depth problems. Because you must record all your calculations in the exam-provided solution booklet to receive full credit, you may want to use graph paper to record your answers for the problems in the depth modules of this book. Depth problems in this book contain three to five subparts, as on the exam. For problems containing subparts, the various parts are referenced as "part 1(a)," "part 1(b)," and so forth, where "1" is the problem number and "(a)" or "(b)" denotes the subpart. Problem subparts are distinguished from solution subparts by the word "problem" (e.g., problem part 1(a)). Occasionally, there may be other acceptable solutions beyond what is presented in this book, and these will be noted in the solution text.

Depth problems and solutions include illustrations that depict structural details, dimensions, and loading information. When multiple illustrations are provided, they are identified by Roman numerals. In problems, these illustrations are referred to as "illustration I," "illustration II," and so forth. When these illustrations are referenced in solutions, they are referred to as "problem illustration I," "problem illustration II," and so forth. Sometimes, a section is cut or a detail is magnified and shown in a subsequent illustration. When this is the case, the illustration uses an arrow to indicate the location of the subsequent illustration, along with text stating "illustration I," "illustration II," and so forth.

A step-by-step solution is provided for each problem. Some solutions include more than one solving method so you can see alternative solution approaches. The text, calculations, and illustrations in the depth solutions are color-coded to indicate which elements must be recorded in your examination booklet to receive full credit on the depth exams. All items appearing in blue are required elements of a complete solution. Items in black are explanatory and included in this book to illustrate the complete solution approach and to increase your exposure to exam-adopted codes. All final answers are presented in boxes for clarity.

When codes are referenced in either the breadth or the depth modules, they are generally referenced by abbreviation. Abbreviations are identified in the "Codes and References" section of this book, preceding their full title.

The nomenclature section lists the variables, symbols, and subscripts used in this book. Whenever possible, variables match those used in a specific code. The use of code-based variables may result in different problems using different variables to represent the same engineering term (e.g., height may be represented as either h or H). When this book's application of a variable differs from the code, a note identifies the difference.

Table 1 *NCEES Structural Engineering (SE) Exam Component Module Specifications*

Friday: vertical forces (gravity/other) and incidental lateral forces component	
morning breadth module	analysis of structures (32.5%)
4 hours	generation of loads (10%)
40 multiple-choice problems	load distribution and analysis methods (20%)
	design and details of structures (67.5%)
	general structural considerations (7.5%)
	structural systems integration (5%)
	structural steel (12.5%)
	cold-formed steel (2.5%)
	concrete (12.5%)
	wood (10%)
	masonry (7.5%)
	foundations and retaining structures (10%)
afternoon depth module[a]	buildings[b]
4 hours	steel structure (1-hour problem)
essay problems	concrete structure (1-hour problem)
	wood structure (1-hour problem)
	masonry structure (1-hour problem)
	bridges
	concrete superstructure (1-hour problem)
	other elements of bridges (e.g., culverts, abutments, and retaining walls) (1-hour problem)
	steel superstructure (2-hour problem)
Saturday: lateral forces (wind/earthquake) component	
morning breadth module	analysis of structures (37.5%)
4 hours	generation of loads (17.5%)
40 multiple-choice problems	load distribution and analysis methods (20%)
	design and details of structures (62.5%)
	general structural considerations (7.5%)
	structural systems integration (5%)
	structural steel (12.5%)
	cold-formed steel (2.5%)
	concrete (12.5%)
	wood (7.5%)
	masonry (7.5%)
	foundations and retaining structures (7.5%)
afternoon depth module[a]	buildings[c]
4 hours	steel structure (1-hour problem)
essay problems	concrete structure (1-hour problem)
	wood and/or masonry structure (1-hour problem)
	general analysis (e.g., existing structures, secondary structures, nonbuilding structures, and/or computer verification) (1-hour problem)
	bridges
	piers or abutments (1-hour problem)
	foundations (1-hour problem)
	general analysis of seismic structures (e.g., seismic and/or wind) (2-hour problem)

[a]Afternoon sessions focus on a single area of practice. You must choose *either* the buildings or bridges depth module, and you must work the same depth module across both exam components.

[b]At least one problem will contain a multistory building, and at least one problem will contain a foundation.

[c]At least two problems will include seismic content with a seismic design category of D and above. At least one problem will include wind content with a base wind speed of at least 110 mph. Problems may include a multistory building and/or foundation.

Most steel problems in this book (and on the SE exam) can be solved using either allowable stress/strength design (ASD) or load and resistance factor design (LRFD), so solutions for both methods are presented. For wood design, the problems and solutions in this book utilize only the ASD method. Although the exam allows solutions using either ASD or LRFD for many wood problems, ASD solutions are used exclusively in this book to save you time on the exam. (ASD and LRFD for wood design are nearly identical, except for the additional conversion factors required for LRFD.) The masonry design problems in this book are solved using the ASD provisions of TMS 402 Chap. 2 to match current NCEES requirements. All concrete design problems use the strength design method of ACI 318.

HOW TO USE THIS BOOK

The best way to prepare for the exam is by solving problems. The exam day environment is intense, so the more experience you have solving problems in a timed setting, the better. Problems from the *Structural Engineering Reference Manual* and its companion book *Structural Engineering Solved Problems*, as well as problems from other textbooks, are a good starting point. But it is important to practice solving many six-minute multiple-choice and one-hour essay problems like those you will see on exam day.

Upon getting this book, it may be tempting to flip through the problems and solutions immediately to see where you stand before you begin studying, but this approach is not recommended. The only way to simulate the exam situation is to solve problems you've never seen before. That is why you should refrain from reviewing the problems in this book until you have completed, or nearly completed, your study of the subject matter.

Start your review by studying the codes and references (see the section later in this Introduction titled "How to Prepare for the Exam"). Then, set aside four-hour blocks of time for each of the four exam modules, and complete the problems in the allotted time. For the problems that you answer correctly, you can feel confident heading into the exam. For the problems that you solve incorrectly, you can spend some time before the exam to further review related topics.

Alternatively, you can use this book to practice solving problems related to a specific topic after you complete your study of that subject area. For example, after you have completed your study of concrete, you may want to locate the concrete problems in this book and solve them separately from the other problems. The design codes listed in the problem statement will help you to recognize the related topics (e.g., any question that lists ACI 318 as a design code is a good question to challenge your understanding of concrete design).

Bring this book with you on exam day. It is quite possible that you'll encounter a couple questions that are very similar to those in this book, in which case you can save time by referring to this book's solutions.

As a general rule, use your time studying the design codes and references to master the material, and use your time solving practice problems to master your exam-taking techniques. For more exam preparation tips, see the subsection later in this Introduction titled "Exam-Taking Strategies and Tips."

LICENSURE FOR STRUCTURAL ENGINEERS

Although the SE exam is available in most states as a standardized structural engineering licensing exam, the benefits of passing the exam vary from state to state. Many states offer a separate SE license contingent on passing the SE exam and meeting other state-specific requirements. Some of these states have title or practice acts, which require a valid SE license in order to perform certain types of structural engineering work. Other states that offer an SE license have no practice restrictions, in which case an SE license entitles the licensee to use the title of SE, but offers no additional benefits. Consult your state board for more information about state engineering licensing requirements. (Visit PPI's website at **ppi2pass.com/stateboards** for links to all state board websites.)

Structural engineers who pass the SE exam in states that do not offer separate SE licensure can receive a PE license only—the same license that can be obtained by taking an 8-hour exam. As a result, many structural engineers in these states wonder why they should bother taking a two-day exam when a one-day exam will earn them the same license. However, even in these states, structural engineers who pass the SE exam have several advantages over structural engineers who pass the PE exam.

The SE exam is the only NCEES exam that exclusively covers material relevant to a structural engineer; passing the exam is the only way for practicing structural engineers to truly demonstrate competency in their profession. Although the 8-hour civil breadth and structural depth PE exam covers structural engineering topics, the morning module of the exam also covers many subjects beyond the expertise of most structural engineers, including topics from environmental, transportation, and water resources engineering.

Another advantage of taking the SE exam is that passing the exam makes you eligible for SE licensure in the future. Even if your state has not yet adopted SE licensure, by passing the SE exam now, you will be eligible to become a licensed SE as soon as your state's licensing board recognizes separate licensure for structural engineers.

Furthermore, the National Council of Structural Engineers Associations (NCSEA), which advocates for national adoption of the SE license, has established the

Structural Engineering Certification Board (SECB). SECB's goal is to identify and certify professionals who are qualified to practice structural engineering. Passing the SE exam can make you eligible for SECB certification, regardless of whether or not your state offers SE licensure, and can make you more marketable to the public as a structural engineer.

Many states with SE licensure allow you to apply for reciprocity (also commonly called comity or endorsement) upon passing the SE exam, so you can become an SE in those states. To aid in this process, NCEES has established the Model Law Structural Engineer (MLSE) designation. This designation can be added to your NCEES record upon passing the SE exam to expedite the comity licensure process.

No matter what the engineering licensure terms are for your state, the best approach is to stay ahead of the curve, dedicate a couple months to studying, and pass the SE exam. The experience is both challenging and rewarding, as you refresh your knowledge of basic structural engineering principles and gain further understanding of the codes that govern the profession.

REGISTERING FOR THE EXAM

You must pass the Fundamentals of Engineering (FE) exam before taking the SE exam, and you must apply to take the exam through the professional licensing board of the state in which you would like to become licensed. The exam is administered twice a year in April and October, and the deadline to apply is typically four to eight months before the exam date. Each state sets its own requirements for the education and experience that must be met before you can take the exam, so check with your local licensing board before applying.

After your state accepts your application, you must then register with NCEES and the examination service company that will be proctoring the exam in your state. Two to three weeks before the exam, NCEES will notify you that your exam-day documents are available to download from its website.

EXAM FORMAT AND CONTENT

The 16-hour SE exam consists of two 8-hour components: the vertical forces component, offered on Friday, and the lateral forces component, offered on Saturday. For the vertical forces component, most problems relate to the analysis and design of structures due to gravity loads such as dead, live, and snow loads; other problems relate to soil and fluid loads. The lateral forces component relates to wind and earthquake loads and the detailing and analysis of lateral systems. (See Table 1 for SE exam topics and their corresponding percentage of the exam.)

Each 8-hour component contains a 4-hour morning breadth module and a 4-hour afternoon depth module, as outlined in Table 1.

For the afternoon depth module, you can choose to answer either building or bridge problems, but you must make the same choice for the vertical and lateral components. It is important to note that, regardless of whether you choose building or bridge problems for the afternoon modules, the morning modules contain both building and bridge problems (roughly 75% buildings and 25% bridges). Therefore, even if you plan to take the building module, you must still be familiar with the bridge design topics from *AASHTO LRFD Bridge Design Specifications*.

To pass the SE exam and be eligible for the SE license, you must obtain an acceptable score on both the vertical and lateral components. However, it is possible to take the components separately (e.g., take the vertical component in April and take the lateral component in October). Most candidates choose to take the full 16-hour exam in a single weekend. It is also possible to take both components together but pass only one. In this case, within a five-year period you would need to retake and pass only the exam component that you did not pass initially.

Breadth Modules (Multiple-Choice Problems)

Both 4-hour breadth modules contain 40 multiple-choice problems, and each problem has four answer options. Each multiple-choice problem is completely independent. In other words, you will never need to use your answer from one problem to solve a subsequent problem. You will be given an examination booklet with one problem on each page and blank space to perform calculations. All calculations must be performed within the examination booklet. If you write on anything except the examination booklet and answer sheet, your exam will be disqualified, and you will be dismissed from the exam site. An answer sheet is also provided for you to record your answers, and only this answer sheet is graded. No credit is given for calculations written in the examination booklet.

There are two main types of problems in the breadth modules: qualitative (approximately 10–20% of the problems) and quantitative (approximately 80–90% of the problems). Qualitative problems generally do not require calculations. Instead, they may ask about engineering principles and practices; about the particular requirements of a specific design code section; or about how to pick the correct deflected shape, shear, or moment diagram. They may also ask you to identify the most appropriate structural detail.

Quantitative problems require calculations. They may ask you to calculate a force, stress, dimension, or other numerical value by using a general engineering equation or an equation from one of the design codes. The problem statements for quantitative breadth problems typically include applicable design codes, design criteria, simplifying assumptions, and four answer options. Most quantitative problems also include an illustration showing the referenced structural component with dimensions and member sizes. Occasionally, the problem statement will contain more information than is necessary to solve the problem.

Depth Modules (Constructed Response/ Essay Problems)

Both 4-hour depth modules in buildings contain four 1-hour essay problems. (The bridge depth modules, which are not included in this book, include two 1-hour essay problems and one 2-hour essay problem.)

The problem statements for depth problems include applicable design codes, design criteria, problems to be solved, and illustrations. The illustrations show structural details, dimensions, and loading information. As with the breadth problems, the depth problems may provide more information than is actually required to solve the problem correctly.

For each problem, there will be approximately three to five subparts labeled with the problem number and a letter corresponding to the subpart. For example, problem 1 will include subparts (a), (b), (c), and so forth. These subparts may ask you to calculate a load, determine a force, design a member, check the adequacy of a member, or sketch a structural detail. All the subparts are at least loosely related, and occasionally you will need to use a value from a previous subpart.

Unlike the multiple-choice modules, your answers must all be handwritten—there is no answer sheet for you to fill out. Instead, a solution booklet of graph paper will be provided for each problem. Use this booklet to show all your calculations and your final answers.

HOW TO PREPARE FOR THE EXAM

Beginning the task of studying for the SE exam can be daunting. Historically, the pass rate for the exam has only been about 35%–50% for first-time takers, which is considerably lower than the pass rate for 8-hour PE exams. These statistics are not meant to scare you, but rather to motivate you to study. If you are well-prepared, you will feel confident on exam day and will be more likely to pass the exam.

Design Codes and Recommended References

The SE exam is an open-book exam. Therefore, you are allowed to bring references to the exam as long as they are bound and remain bound. Personal notes in a three-ring binder and other semipermanent covers can usually be used. However, some states use a "shake test" to eliminate loose papers from binders, so make sure nothing escapes from your binders when they are inverted and shaken.

A few states maintain a formal list of banned books or don't permit collections of solved problems, sample exams, and solution manuals. Therefore, check with your state's board of engineering registration for any restrictions. (The PPI website has a listing of state boards at **ppi2pass.com/stateboards**.)

The Codes and References section of this book lists exam-adopted design codes and recommended references. As a general rule, if you did not use a book during your preparation, don't bother bringing it on exam day. Most of the problems on the SE exam can be solved using your engineering judgment and the NCEES design codes, although other references may help you to answer a few problems or may be useful during your review.

Creating a Study Schedule and Obtaining Study Materials

The SE exam covers many topics and design codes, so the best strategy is to start studying early. Your total ideal study time depends on your experience, but successful examinees report spending 200–300 hours preparing for the exam. You should begin studying around six months before the exam and dedicate 10–20 hours each week to rigorous study. There is no one correct way to study—it is up to you to determine whether your learning style is best suited to independent study, study groups, webinars, or review courses (or a combination of these methods).

As soon as possible, gather all of your study materials and make sure that they are up to date.

1. Check the NCEES website for the current exam specifications and exam-adopted design codes.

2. Obtain copies of all the exam-adopted design codes. (The Codes and References section of this book includes the codes adopted at the time of publication and other books you may need.) Be sure that you have a copy of the correct edition of the listed code. Some design codes don't change significantly between editions, but it is possible to solve a problem incorrectly simply by using the wrong edition. Buying all of these books can be expensive, so it is worthwhile to see if your employer or colleagues have copies that you can borrow.

3. Review any errata and amendments to design codes and mark them in your books.

In addition to obtaining the reference books, you must use an NCEES-approved calculator. Visit PPI's website at **ppi2pass.com/calculators** for a complete list of approved calculators. The best choice is a model with multiple lines of text display, which allows you to see the equations you have typed and to check that you haven't made an input error. To maximize your speed on exam day, use this calculator throughout your studying and also during your daily work.

The next step is to set aside blocks of time every week when you will devote yourself to following a detailed study schedule. An ideal study schedule would include two- to five-hour sessions at least two to three times a week. (If you find it difficult to focus in the evenings after a full day of work, try studying mostly in the mornings and on weekends.) It is especially important to create a detailed study schedule if you are among the majority of examinees who are employed full time and have limited availability for studying. Make a list of the topics you want to study each week, and stick to it. Try to spend 25–50 hours on each of the following topics (the topics that are most prevalent on the exam are listed first).

- loads (ASCE/SEI7 and IBC)

- general structural analysis, including seismic design (IBC, textbook on structural analysis, and other references)

- concrete (ACI 318 and *PCI Design Handbook*)

- steel (AISC *Steel Construction Manual* and AISC *Seismic Design Manual*)

- bridges (*AASHTO LRFD Bridge Design Specifications*)

- foundations (textbook on foundations)

- wood (NDS *National Design Specification for Wood Construction ASD/LRFD*)

- masonry (TMS 402/602 *Building Code Requirements and Specification for Masonry Structures*)

To help you determine which sections of the design codes are most relevant, refer to the SE exam specifications listed in Table 1. It is a good idea to start by reviewing the topics you are least familiar with, as you will have the most difficulty retaining this material. Then, return to the subjects you are least familiar with a month or so before the exam.

Supplement your design code review by reading related sections of the *Structural Engineering Reference Manual* and other textbooks to understand how the code sections are applied in design. The Codes and References section of this book lists additional resources that you may find helpful to your studies.

When determining how to spend these hundreds of hours studying, it is important to keep in mind that this is an open-book exam. There is no need to commit any code text or equations to memory. More than anything else, the exam is about being able to find and apply the relevant design code sections quickly and correctly. It is a good idea to study the design codes carefully to ensure that you understand the provisions and that you are able to navigate through them. When studying tables and figures, read all of the footnotes—the fine print is fair game for an exam problem.

The following are good practices to follow when preparing your references for studying and exam day.

- Check your state's exam requirements and restrictions, as some states restrict which books and materials can be used for the exam. (The PPI website has a listing of state boards at **ppi2pass.com/stateboards**.)

- Place tabs with neatly written labels on the pages of your design codes and references to help you quickly find important sections and equations during the exam. Tabs must be permanent. Loose pieces of paper or sticky notes that are easy to remove aren't permitted.

- Write notes in the margins of the design codes that help you to better understand the provisions or that direct you to other relevant code sections. (However, be very careful to not write anything in your references during the exam—this is strict grounds for your exam to be disqualified and for you to be dismissed from the exam site.)

- Create personal summary sheets of the most important equations and design code procedures. For example, a seismic summary sheet might include the step-by-step procedure for determining the base shear of a building. Summary sheets serve a dual purpose, as they help you to better understand the code provisions as you write them down and, if allowed by your state, function as an easy-to-use reference on exam day. Be sure to bind the sheets (using a three-ring binder, spiral binding, etc.) before bringing them to the exam. Stapled pages and notepads aren't permitted for use on the exam by any states.

- Compile a set of typical details as you study, as the depth problems will often require you to sketch structural details. If your state allows bound personal notes or photocopy drawings of connection and reinforcing details, put them in a binder organized by material type so you can easily reference the details during the exam.

Exam-Taking Strategies and Tips

To pass the SE exam, there is no substitute for studying, studying, and studying some more. However, there are some basic strategies that you can employ on exam day to help you maximize your score.

Keep track of time. For the breadth modules, you have an average of only six minutes to solve each problem, so periodically check the clock to make sure you are on track to complete about ten problems per hour. If you've spent more than five minutes on a problem and aren't getting close to an answer, move on and come back to that problem at the end. Similarly, for the depth modules, don't spend much more than one hour on a single problem—your accuracy will slip if you end up rushing through the remainder of the problems. If you are struggling with one depth module subpart, make an assumption that allows you to move on to the next subpart.

Mark an answer for each breadth problem. For the breadth modules, there is no penalty for incorrect answers, so it is imperative that you mark an answer for every problem. Ideally, you'll want to maintain a pace that gives you 15 minutes at the end to return to the problems you found tricky and check that the bubbles on your answer sheet correspond with your intended answers. For any calculation-heavy problems, reenter the equations into your calculator to check that you did not make an input error. During the last five minutes, make sure all of the bubbles are dark and fully filled in, and make a guess for any problems you weren't able to attempt.

Read problem statements carefully. While the problems on the exam are mostly straightforward because they are designed to determine your competency rather than "trick" you into giving a wrong answer, some problems may include a key piece of information that is easy to overlook. Some examples of details that are easy to overlook include

- the problem specifying lightweight concrete

- the steel reinforcement having a strength of $F_y = 40$ kips/in^2

- the problem specifying the loads in kips but asking for an answer in tons

It is easy to miss these types of details if you are rushing while reading the problem statement. Time spent carefully reading each problem is time well spent.

For the depth problems, be sure to read the entire problem statement before beginning your solution. An illustration on the next page may include valuable information.

Understand what you should be solving for. Pay close attention to exactly what the problem is asking you to calculate. For example, if the problem asks for the "nominal strength" or "nominal capacity," do not include any ϕ, Ω, or other adjustment factors in your calculation. However, if the problem asks for the "design strength," "allowable strength," or "adjusted design value," do include ϕ, Ω, or other adjustment factors as appropriate. If the problem asks you to calculate a load, force, or stress, you need to determine whether you should be using factored or unfactored loads; generally, if the problem asks for a "design" force, you should include appropriate load factors, and if the problem asks for a "service level" force, you should use unfactored loads.

Know what "most nearly" means. The term "most nearly" can be a little tricky: NCEES uses this phrase for quantitative problems to account for the fact that

many problems will have a range of correct answers. This range of answers may result from the rounding of intermediate values or small differences in assumptions (e.g., assuming a self-weight for normal weight concrete of 145 lbf/ft^3 instead of 150 lbf/ft^3). If your answer falls halfway between two of the choices, this may be a sign that you have made a mistake. However, it is normal for an answer not to match one of the listed choices exactly, even if you have precisely solved the problem without intermediate rounding or assumptions of any kind. Do not spend valuable exam time trying to calculate an exact match. Mark the answer that is most nearly correct, and then move on to the next problem. Also ensure that you are rounding up or down appropriately. For example, if the problem asks for the maximum load a beam can carry, you must round your calculated value down (as rounding up would exceed the maximum capacity).

Ace easy problems. There will be a couple of problems in the breadth module that may seem a little too easy: don't overcomplicate them. By getting these problems correct and moving on, you will have more time to spend on the difficult problems.

Know some shortcuts. There are many ways to simplify the code equations to help you save time on exam day. For example, rather than solve a quadratic equation, you can use the expression $A_s = M_u/4d$ to approximate the required area of flexural steel for tension-controlled concrete members. In addition, it is common for code sections to list several equations where only the largest or smallest calculated value is applicable. Therefore, it is a good idea to write in the margins of the code when each equation controls. If you know this ahead of time, you can minimize the number of calculations you need to perform on exam day.

Adjust the order you solve problems in as appropriate. The breadth module question order roughly follows the topic order listed in the exam specifications (see Table 1). However, questions that cover similar topics are not always grouped together. As a result, some examinees find it advantageous to solve the breadth problems out of order by grouping together problems with the same design code listed in the "Design Criteria." For example, you may want to solve the bridge problems consecutively to avoid having to take out and put away the bulky AASHTO code several times during the exam. If you choose to solve the problems out of order, use extra care to ensure that the bubbles you fill in on your answer sheet correspond to the correct problems.

Use indexes. When you come across a problem that you have no idea how to begin, the best approach is to look in the index or table of contents of the referenced design code. Often, this can help you to find the relevant code section more quickly, especially in the massive AASHTO code.

Look at the answer options before starting to solve the problem. The options given for the breadth problems can often provide valuable insight into how to approach a problem. For example, if the answers are very close together, you'll have to be precise in your calculations, whereas if there is a wider range of answers, you can simplify calculations, take shortcuts, or round intermediate values. Also, it may occasionally make sense to work backward. For example, if you are trying to determine the lightest adequate shape for a steel column, it may be faster just to look up the tabulated capacities of the four answer options rather than try to solve for the most efficient shape directly.

Validate your answers. The quantitative breadth problems are often designed so that the incorrect answers can be arrived at by making common mistakes. If you recognize what some common mistakes might be, you can analyze whether your answer is more likely to be correct by quickly checking to see if common mistakes lead to the other answer options. Additionally, for steel and wood problems, you may have the option to solve the problem using either ASD or LRFD. If time allows, it is a good idea to check your answer by solving the problem both ways to make sure you still arrive at the same answer option.

Have a good understanding of what graders are looking for. For the breadth problems, the grader is a machine, and what the machine is looking for is very simple: a dark circle filled in on your answer sheet that corresponds to the correct answer. For the depth problems, the grader is a person, and what he or she is looking for is not as obvious. NCEES does not publish any information about how the depth problems are graded, but it can reasonably be assumed that the graders are looking for you to display sound engineering judgment and competency in the subject matter. Getting the correct numerical answers is important, but it is more important that you use the correct code provisions and don't skip any applicable code checks.

With this in mind, you should always

- reference the exact code section you are using by adding the section number in either the text of your answer or the right margin

- state clearly if you skip a check because you think it is not required by inspection

- clearly note any assumptions that were made and why they are appropriate

- write as neatly as you can while still keeping the time limit in mind (although you are not graded on your penmanship, a neat solution conveys a better overall impression to the grader)

- start each equation on a new line and maintain a uniform margin on the left

- use a straight edge and label everything you think might be relevant for sketches

- describe in words how you would answer the problem. If you run out of time, list the code sections you would use if you had enough time to complete the problem.

Before Exam Day

Spend the last week or two before the exam making sure you can quickly navigate the codes using your tabs. Solve as many practice problems as possible to make sure you are up to speed.

If possible, visit the exam site before exam day. Figure out the best way to get there and where to park. If you live far from the exam site, consider reserving a room at a nearby hotel for the night before the exam.

The SE exam is open-book, so try to resist the urge to cram on the night before the exam. Instead, relax and get as much sleep as possible.

Exam Day

On the morning of the exam, eat a high-protein breakfast, and give yourself plenty of time to get to the site.

Getting all your books to the exam site can be a challenge. Consider using a large suitcase with wheels, and have a plan for organizing your books on exam day. Most likely, the provided desk will not be large enough for all your references. One way to keep your books organized and within reach is to place them in milk crates on the floor.

After completing the vertical forces component of the exam on Friday, try not to think too much about how you performed. Relax, get some sleep, and prepare to do it all again on Saturday.

What You Will Need on Exam Day

You should bring the following items with you to the exam.

- A letter of admittance

- A current, signed, government-issued photo ID (not a student ID card)

- An NCEES-approved calculator

- A spare calculator (you can likely borrow one from a coworker for the weekend)

- A straight edge for drawing sketches and reading alignment charts (a 6 in ruler works best)

- A wristwatch or small desk clock: There will be an official clock in the room, but it may not be visible from your seat.

- Earplugs

- Several layers of clothing: Be prepared for the exam room to be hot or cold.

- A water bottle: Confirm with your state that drinks are allowed.
- Approved reference materials

Certain items, such as cell phones and your own pens or pencils, for example, should be left in your test room locker or car. Check with your state board (**ppi2pass.com/ stateboards**) to confirm what is and is not allowed in the exam room.

AFTER THE EXAM

If you think any problems are flawed, ask the proctor for a reporting form immediately after the exam. Follow the proctor's instructions for preparing and submitting the form. The problems on the exam are thoroughly reviewed and errors are rare, but they can occur.

After leaving the exam room, try not to think too much about the problems or how you did. Know that it is normal to get tripped up on some problems and that it is not worth stressing about. Enjoy your newfound free time, and apply what you've learned to your projects at work.

Exam Scoring

NCEES typically releases scores to state licensing boards two to three months after the exam (sometime in December or January if you took the exam in October and sometime in June or July if you took the exam in April). Because some state boards choose not to release the results right away, you might find out your results on a different day than someone else who took the same exam.

The exam is scored on a pass or fail basis, and only those who fail receive detailed results. The exam is neither curved such that a certain percentage of people pass each time nor is there a consistent cutoff score that you need to meet to pass. Instead, NCEES works with subject experts and statisticians to determine the score that corresponds with a minimum competency level: this becomes the passing score. The passing score is not published, and it varies based on the difficulty of each exam. However, based on scores reported by examinees who did not pass, it is believed that the passing score has historically been around 70%.

Although NCEES grades the breadth problems as simply "correct" or "incorrect," its system for grading depth problems is not transparent. Each depth problem receives a grade of "unacceptable," "needs improvement," or "acceptable," but it is unknown what the criteria are for each category or how the depth scores are combined with the breadth scores. Suffice it to say that you will need to do well on both the breadth and depth modules to receive a passing score.

After You Receive Your Results

If you fail, do not be discouraged. Historically, the pass rate for the SE exam has been low. The studying you have already done has not gone to waste: your code books are still tagged, and everything you have learned can easily be refreshed with brief study sessions. Remember that if you pass only the vertical or lateral component of the exam, you need to retake and pass only the other component within five years to be eligible for licensure. If you do not pass either component, consider studying for and taking only one 8-hour component at a time. Use the diagnostic report that you received from NCEES to figure out which subjects you need to study more. Dedicate more time for review, solve practice problems, and then take the exam again.

Once you pass, it is time to celebrate. Fill out any paperwork required to get your license, thank everyone who supported you while you prepared, and ask your employer for a raise!

Codes and References

The minimum recommended library for the SE exam includes the NCEES exam-adopted design codes and the *Structural Engineering Reference Manual.* You do not need an extensive library beyond these books on exam day. Most problems on the SE exam can be solved using the exam-adopted design codes and your knowledge of general engineering principles. As a general rule, you shouldn't bring books to the exam that you didn't use during your exam review. This section divides useful materials for SE exam preparation into three categories: NCEES Codes, Additional Recommended References for Exam Day, and Additional Recommended References for Exam Preparation. Codes and references in this section followed by an asterisk (*) were used in the preparation of this book.

The information that was used to write this book was based on exam specifications and adopted design codes at the time of publication. However, as with engineering practice itself, the SE exam is not always based on the most current codes or cutting-edge technology. Similarly, codes, standards, and regulations adopted by state and local agencies often lag issuance by several years. It is likely that the codes you use in practice and the codes that are the basis of your exam will be different.

PPI lists on its website the dates and editions of the codes, standards, and regulations on which NCEES has announced that the SE examination is based (**ppi2pass.com/structural**). It is your responsibility to find out which codes are relevant to the SE exam. In the meantime, here are the codes that have been incorporated into this edition.

NCEES CODES[1]

AASHTO: *AASHTO LRFD Bridge Design Specifications,* Seventh ed., 2015. American Association of State Highway and Transportation Officials, Washington, DC.*

ACI 318: *Building Code Requirements for Structural Concrete,* 2014. American Concrete Institute, Farmington Hills, MI.*

AISC: *Seismic Design Manual,* Second ed., 2012. American Institute of Steel Construction, Inc., Chicago, IL.* (Lateral forces exam only.)

AISC: *Steel Construction Manual,* Fourteenth ed., 2011. American Institute of Steel Construction, Inc., Chicago, IL.*

AISI S100: *North American Specification for the Design of Cold-Formed Steel Structural Members,* 2012 edition. American Iron and Steel Institute, Washington, DC.*

AISI S213: *North American Standard for Cold-Formed Steel Framing—Lateral Design 2007 Edition with Supplement No.1,* October 2009 (reaffirmed 2012). American Iron and Steel Institute, Washington, DC.

ASCE/SEI7: *Minimum Design Loads for Buildings and Other Structures,* 2010, 3rd printing. American Society of Civil Engineers, Reston, VA.*

IBC: *International Building Code* (without supplements), 2015. International Code Council, Falls Church, VA.*

NDS: *National Design Specification for Wood Construction ASD/LRFD,* 2015. *National Design Specification Supplement: Design Values for Wood Construction,* 2015. American Forest & Paper Association, Washington, DC.*

NDS: *Special Design Provisions for Wind and Seismic,* 2015 ed. American Wood Council, Leesburg, VA.

TMS 402/602 (ACI 530/530.1)[2]: *Building Code Requirements and Specification for Masonry Structures* (and related commentaries), 2013. The Masonry Society, Boulder, CO; American Concrete Institute, Detroit, MI; and Structural Engineering Institute of the American Society of Civil Engineers, Reston VA.*

ADDITIONAL RECOMMENDED REFERENCES FOR EXAM DAY

American Institute of Steel Construction, Inc. *AISC Design Guide 11: Floor Vibrations Due to Human Activity.* Chicago, IL.*

---

[1]These codes and standards apply to the vertical and lateral components of the SE exam. Solutions to exam problems that reference a code or standard are scored based on this list. Solutions based on editions or codes or standards other than those specified by NCEES at the time of your exam will not receive credit.
[2]TMS 402 is also known as ACI 530 or ASCE 5.

American Institute of Steel Construction, Inc.[3] *Prequalified Connections for Special and Intermediate Steel Moment Frames for Seismic Applications*, including Supplement no. 1, 2010. Chicago, IL.* (Lateral forces exam only.)

American Institute of Steel Construction, Inc.[4] *Seismic Provisions for Structural Steel Buildings*, 2010 ed. Chicago, IL. (Lateral forces exam only.)

American Institute of Steel Construction, Inc.[5] *Specification for Structural Steel Buildings*, 2010 ed. Chicago, IL.*

Hibbeler, R. C. *Structural Analysis*. Upper Saddle River, NJ: Pearson/Prentice Hall. (Or a similar structural analysis textbook.)

National Concrete Masonry Association. *Section Properties of Concrete Masonry Walls* (NCMA TEK 14-1A). Herndon, VA: National Concrete Masonry Association. (This resource is available online as a free download.)

The National Council of Examiners for Engineering and Surveying. *Structural Engineering: Sample Questions + Solutions*. Clemson, SC: The National Council of Examiners for Engineering and Surveying.

Precast/Prestressed Concrete Institute. *PCI Design Handbook: Precast and Prestressed Concrete*, Seventh ed., 2010. Chicago, IL.*

Structural Engineers Association of California. *2015 IBC SEAOC Structural/Seismic Design Manual*, Volume 1, Code Application Examples. Sacramento, CA: Structural Engineers Association of California. (Lateral forces exam only.)

Williams, Alan. *Structural Engineering Reference Manual*. Belmont, CA: Professional Publications, Inc.*

ADDITIONAL RECOMMENDED REFERENCES FOR EXAM PREPARATION

ACI Committee 340. *ACI Design Handbook*. Detroit, MI: American Concrete Institute.*

American Institute of Steel Construction. AISC *Design Examples*. New York, NY: American Institute of Steel Construction. (This resource is available online as a free download.)

American Institute of Steel Construction. *Detailing for Steel Construction*. Chicago, IL: American Institute of Steel Construction.

American Iron and Steel Institute. *Cold-Formed Steel Design: AISI Manual*. Washington, DC: American Iron and Steel Institute.*

Amrhein, James E. *Reinforced Masonry Engineering Handbook*. Washington, DC: Masonry Institute of America. (Or a similar masonry textbook.)

ASTM C42/C42M-16, *Standard Test Method for Obtaining and Testing Drilled Cores and Sawed Beams of Concrete*. West Conshohocken, PA: ASTM International.

Baradar, Majid. *Seismic Design Solved Problems*. Belmont, CA: Professional Publications, Inc. (Lateral forces exam only.)

Breyer, Donald E. and John A. Ank. *Design of Wood Structures*. New York, NY: McGraw-Hill Companies, Inc. (Or a similar wood textbook.)

Building Seismic Safety Council for the Federal Emergency Management Agency of the Department of Homeland Security. *NEHRP Recommended Provisions: Design Examples* (FEMA 451). Washington, DC: U.S. National Institute of Building Sciences. (This resource is available online as a free download.)

Das, Braja M. *Principles of Foundation Engineering*. Boston, MA: PWS-Kent Pub Co. (Or a similar foundation engineering textbook.)

Lindeburg, Michael R. with Kurt McMullin. *Seismic Design of Building Structures*. Belmont, CA: Professional Publications, Inc. (Lateral forces exam only.)

MacGregor, James. *Reinforced Concrete: Mechanics and Design*. Upper Saddle River, NJ: Pearson/Prentice Hall. (Or a similar concrete textbook.)

Naeim, Farzad. *The Seismic Design Handbook*. New York, NY: Van Nostrand Reinhold.*

Salmon, Charles G. and John Edwin Johnson. *Steel Structures: Design and Behavior*. Upper Saddle River, NJ: Pearson/Prentice Hall. (Or a similar steel textbook.)

Structural Engineers Association of California. *2015 IBC SEAOC Structural/Seismic Design Manual*, Volume 2, Examples for Light-Frame, Tilt Up, and Masonry Buildings. Sacramento, CA: Structural Engineers Association of California. (Lateral forces exam only.)

Structural Engineers Association of California. *2015 IBC SEAOC Structural/Seismic Design Manual*, Volume 3, Examples for Concrete Buildings. Sacramento, CA: Structural Engineers Association of California. (Lateral forces exam only.)

Structural Engineers Association of California. *2015 IBC SEAOC Structural/Seismic Design Manual*, Volume 4, Examples for Steel-Framed Buildings.

---

[3]AISC 358-10 is found in Part 9.1 of the AISC *Seismic Construction Manual*, Second ed.
[4]AISC 341-10 is found in Part 9.1 of the AISC *Seismic Construction Manual*, Second ed.
[5]AISC 360 is found in the AISC *Steel Construction Manual*, Fourteenth ed.

Sacramento, CA: Structural Engineers Association of California. (Lateral forces exam only.)

Structures and Codes Institute. *CodeMaster—Seismic Design.*[6] Palatine, IL: Structures and Codes Institute: S. K. Ghosh Associates, Inc. (Lateral forces exam only.)

Structures and Codes Institute. *CodeMaster—Wind Design Overview.* Palatine, IL: Structures and Codes Institute: S. K. Ghosh Associates, Inc. (Lateral forces exam only.)

Wheat, Dan L., Steven M. Cramer, American Forest & Paper Association, and American Wood Council. *ASD/ LRFD Examples: Structural Wood Design Solved Example Problems.* Washington, DC: American Forest & Paper Association.

---

[6] *CodeMaster* guides are laminated code summary sheets. *CodeMaster—Seismic Design* and *CodeMaster—Wind Design* Overview both refer to IBC (2015) and ASCE/SEI7 (2010).

Nomenclature

a	dimension	in
a	geometric parameter for eccentrically loaded weld groups	–
a	longitudinal spacing of moving concentrated loads	in
A	anchor	in^2
A	area	in^2
A	fatigue detail category constant	(lbf/in^2)^3
A_b	distributed bars	in^2
A_c	area of concrete core	in^2
A_c	area of concrete in composite slab	in^2
A_c	area of concrete section resisting shear transfer (ACI 318)	in^2
$A_{c,v}$	area of concrete section resisting shear transfer (AASHTO)	in^2
A_{cv}	gross area of concrete section	in^2
A_{cw}	area of concrete section of a wall or coupling beam resisting shear	in^2
A_{lw}	link web area (excluding flanges)	in^2
A_{nv}	net shear area	in^2
A_{pt}	projected tension area on masonry surface of a right circular cone	in^2
A_s	adjusted peak seismic ground acceleration at short period (AASHTO)	–
A_s	area of nonprestressed longitudinal tension flexural reinforcement	in^2
A_{sv}	area of vertical shear reinforcement	in^2
A_{ts}	area of nonprestressed reinforcement in a tie	in^2
A_v	(required) area of shear reinforcement	in^2
A_{vd}	area of reinforcement in a group of diagonal bars of a coupling beam	in^2
A_{vf}	area of reinforcement for interface shear	in^2
A_{Vc}	project concrete failure area of a single anchor	in^2
A_{Vco}	projected concrete failure area of a single anchor not limited by edge distance, spacing, or member thickness	in^2
ADTT	number of trucks per day in one direction averaged over bridge design life	–
ADTT$_{SL}$	number of trucks per day in a single lane averaged over bridge design life	–
b	dimension	in
b_1	dimension of critical section b_o measured in direction of span	in
b_1	dimension of critical section b_o measured in direction perpendicular to span	in
b_o	perimeter of critical section for shear in concrete slabs and footings	in
b_s	effective width of concrete deck	in
b_s	width of concrete girder	in

b_{vi}	interface width considered to be engaged in shear transfer	in
b_w	width of concrete or masonry section	in
B	base plate dimension parallel to column flange	in
B	column spacing	in
B	distance between diaphragm straps	in
B	footing dimension	in
B	width	in
B_1	moment amplifier to account for second order effects caused by displacements between brace points	–
B_a	allowable axial tensile load of headed anchor bolts	lbf
B_{ab}	allowable tensile load on anchor bolt governed by masonry breakout	lbf
B_{as}	allowable tensile load on anchor bolt governed by steel yielding	lbf
c	cohesion	lbf/in^2
c	depth of cut at center of a reduced beam section	in
c	distance from centroidal axis of critical section to perimeter of critical section	in
c	length	in
c	wood column stability parameter	–
c_1	column dimension in direction of span	in
c_2	column dimension in direction perpendicular to span	in
c_{a1}	distance from center of anchor to edge of concrete in direction of applied shear	in
c_{AB}	distance to neutral axis (edge column)	in
c_b	development length parameter	in
c_c	clear cover of reinforcement	in
c_{top}	distance from the centroid of the reinforced section to the top fiber	in
C	bearing factor	–
C	coefficient for eccentrically loaded weld groups	–
C	compression force	lbf
C_1	electrode strength coefficient	–
C_b	bearing area factor	–
C_b	lateral-torsional buckling modification factor	–
C_d	deflection amplification factor	–
C_D	load duration factor	–
C_{di}	diaphragm factor for nailed connections	–
C_e	exposure factor	–
C_{eg}	end grain factor for wood connections	–
C_F	size factor for sawn lumber	–
C_g	group action factor for wood connections	–
C_h	web slenderness coefficient	–
C_I	incising adjustment factor for dimension lumber	–

C_L	lateral drag coefficient	–
C_m	factor relating actual moment diagram to an equivalent, uniform moment diagram	–
C_M	wet service adjustment factor	–
C_N	bearing length coefficient	–
C_{NW}	net windward pressure coefficient	–
C_P	column stability factor	–
C_{pr}	factor to account for peak connection strength	–
C_R	inside bend radius coefficient	–
C_s	seismic response coefficient for buildings	–
C_s	slope factor for sloped roof snow loads	–
C_{SG}	adjusted specific gravity	–
C_{sm}	seismic response coefficient for bridges	–
C_t	approximate building period parameter	–
C_t	temperature adjustment factor	–
C_T	buckling stiffness factor for dimension lumber	–
C_{tn}	toe-nail factor for nailed connections	–
C_{vx}	base shear vertical distribution factor at building level x	–
C_Δ	geometry factor for wood connections	–
d	depth of section (steel and wood)	in
d	diameter	in
d	distance	in
d	effective depth to tension reinforcement (concrete and masonry)	in
d	thickness	in
d'	distance	in
d'	effective depth to compression reinforcement	in
d_b	nominal diameter of reinforcing bar	in
d_h	width of bolt hole considered in net area calculation	in
d_x	distance from centroid of pile group to centroid of individual pile about the x-axis	in
d_y	distance from the centerline of the pile group to a transverse row of piles or half the longitudinal pile spacing	in
D	dead load	lbf, lbf/in, lbf/in^2
D	diameter	in
D	superstructure depth	in
D	weld size	sixteenths-of-an-inch
D_p	distance from top of concrete deck to plastic neutral axis	in
e	eccentricity	in
e	length of link in an eccentrically braced frame	in
e_g	distance between centers of gravity of beam and deck	in
e_x	eccentricity of the load	in
E	earthquake load	lbf/in, lbf/in^2
E	length of tapered tip of wood screw	in
E	modulus of elasticity	lbf/in^2
E	seismic effect	lbf
EAF	exposure adjustment factor	–
E_{min}	reference modulus of elasticity for beam stability and column stability calculations	lbf/in^2

E'_{min}	adjusted modulus of elasticity for beam stability and column stability calculations	lbf/in^2
EI	flexural stiffness of compression member	in^2-lbf
f	stress	lbf/in^2
f_1	live load adjustment factor	–
f_2	snow load adjustment factor	–
f_a	calculated compressive stress in masonry due to axial loads	lbf/in^2
f_b	calculated compressive stress in masonry due to flexure	lbf/in^2
f_b	maximum flexural/compression stress	lbf/in^2
f_{b1}	maximum bending stress in existing beam prior to installation of reinforcement	lbf/in^2
f_{b2}	maximum bending stress of top of composite section	lbf/in^2
f'_c	specified compressive strength of concrete	lbf/in^2
f'_m	specified compressive strength of masonry	lbf/in^2
f_{pu}	specified tensile strength of prestressing steel	lbf/in^2
f_s	calculated reinforcement stress	lbf/in^2
f_t	bending stress at top of beam	lbf/in^2
f_v	calculated shear stress in masonry	lbf/in^2
f_y	specified yield strength of steel reinforcement	lbf/in^2
f_{yh}	specified yield strength of transverse reinforcement (AASHTO)	lbf/in^2
f_{yt}	specified yield strength of transverse reinforcement (ACI 318)	lbf/in^2
f_z	factor to account for increase in ice thickness with height	–
f_θ	calculated bearing stress for a member bearing at an angle to grain	lbf/in^2
F	force or applied load	lbf
F	net horizontal force/restorative force	lbf
F'	horizontal force	lbf
F_a	short-period site coefficient	–
F_b	allowable compressive stress in masonry	lbf/in^2
F_b	reference bending design value for wood	lbf/in^2
F'_b	adjusted bending design value for wood	lbf/in^2
F_c	reference compression design value parallel to grain	lbf/in^2
F_{cE}	critical buckling design value for wood compression members	lbf/in^2
F_{cr}	elastic critical buckling stress	lbf/in^2
F_{cre}	critical buckling stress using expected yield stress	lbf/in^2
F'_c	adjusted compression design value parallel to grain	lbf/in^2
F_c^*	reference compression design value multiplied by all applicable adjustment factors except C_p	lbf/in^2
$F_{c\perp}$	reference compression design value perpendicular to grain	lbf/in^2
$F'_{c\perp}$	adjusted compression design value perpendicular to grain	lbf/in^2
F_{EXX}	welding electrode classification number	lbf/in^2
F_I	portion of seismic base shear at building level I	lbf

Symbol	Description	Units
F_n	horizontal seismic force at floor n	lbf
F_p	anchorage design force	lbf
F_p	net horizontal restorative force due to passive soil resistance	lbf
$F_{\text{res,tot}}$	total sliding resistance	lbf/in
F_s	allowable stress in reinforcement	lbf/in²
F_t	reference tensile design value parallel to grain	lbf/in²
F_t'	reference tensile design value parallel to grain	lbf/in²
F_u	factored tensile force	lbf
F_u	specified tensile strength	lbf/in²
F_v	allowable shear stress	lbf/in²
F_{vm}	allowable shear stress resisted by masonry	lbf/in²
F_v	long-period site coefficient	–
F_V	reference shear design value parallel to grain for wood	lbf/in²
F_V	transverse force at top of pier	lbf
F_y	longitudinal force at top of pier	lbf
F_y	specified yield strength of steel	lbf/in²
F_{yb}	base plate steel	lbf/in²
F_{yc}	column steel	lbf/in²
F_z	vertical load	lbf
F_θ'	adjusted bearing design value for a member bearing at an angle to grain	lbf/in²
$(\text{FS})_{\text{sliding}}$	factor of safety for sliding of a foundation	lbf
g	acceleration of gravity, 32.2	ft/sec²
g	transverse center-to-center spacing (gage) between fastener gage lines	in
G	gust-effect factor	–
G	shear modulus of elasticity	lbf/in²
G_A	alignment chart factor	–
G_B	alignment chart factor	–
GC_p	product of external pressure coefficient and gust-effect factor	–
GC_{pf}	product of equivalent external pressure coefficient and gust-effect factor	–
GC_{pi}	internal pressure coefficient	–
h	depth of flat portion of web of cold-formed steel shape	in
h	height	in
h_a	thickness of member in which an anchor is located	in
H	height	in
H	load due to lateral earth pressures, ground water pressure or pressure of bulk materials	lbf
I	importance factor	–
I	moment of inertia	in⁴
I_e	seismic importance factor	–
I_I	ice importance factor—thickness	–
I_s	snow importance factor	–
$I_{x,e}$	moment of inertia of existing beam	in⁴
I_{xx}	moment of intertia taken around x-x axis	in⁴
IM	dynamic load allowance	–
j	lever-arm factor for elastic design of reinforced masonry	–
J	polar moment of inertia	in⁴
k	base shear distribution exponent	–
k	effective length factor for compression member	–
k	exponent related to the structure period	–
k	geometric parameter for eccentrically loaded weld groups	–
k	neutral axis depth factor for elastic design of reinforced masonry	–
k	seismic acceleration coefficient	–
K	bridge lateral stiffness	lbf/in
K	dimension used to calculate reinforcement development in masonry	in
K	effective length factor	–
K_1	fraction of concrete strength available to resist interface shear	–
K_1	multiplier for topographic wind factor	–
K_2	limiting interface shear resistance	lbf/in²
K_2	multiplier for topographic wind factor	–
K_3	multiplier for topographic wind factor	–
K_a	coefficient of active soil pressure	–
K_{AE}	seismic active pressure coefficient	–
K_{cr}	time dependent deformation (creep) factor for deflection of wood beams	–
K_d	wind directionality factor	–
K_e	buckling length coefficient for wood compression members	–
K_g	longitudinal stiffness parameter	in⁴
K_{LL}	live load element factor for live load reduction	–
K_o	coefficient of at-rest soil pressure	–
K_p	coefficient of passive soil pressure	–
K_{tr}	transverse reinforcement index	–
K_x	effective column length for bending about strong axis	–
K_y	effective column length for bending about weak axis	–
K_z	velocity pressure exposure coefficient for determination of wind loads at height z	–
K_{zt}	topographical factor for determination of wind loads	–
KL	effective unbraced length	in
l	critical base plate cantilever dimension	in
l	length	in
l_1	length of span in direction that moments are being determined	in
l_1	total bar length for heel bars	in
l_2	length of span in direction perpendicular to l_1	in
l_2	total bar length for toe bars	in
l_b	effective embedment length of headed or bent anchor bolt	in
l_d	development length of reinforcement	in
l_{dh}	development length of reinforcement with a standard hook	in
l_n	length of clear span measured face-to-face of supports	in
l_u	length of roof upwind of snow drift	in
L	bridge span	in
L	lateral load	lbf, lbf/in, lbf/in²
L	length	in
L	live load	lbf, lbf/in, lbf/in²
L	superstructure span	in

L_b	length between points braced against lateral displacement of compression flange or braced against twist of the cross section	in
L_c	clear distance between edge of hole and edge of adjacent hole	in
L_h	distance upwind of crest of escarpment to where the difference in ground elevation is half the height of escarpment	in
L_o	unreduced design live load	lbf/in^2
L_p	limiting laterally unbraced length for the limit state of yielding	in
L_{pd}	limiting laterally unbraced length for plastic analysis	in
L_r	limiting laterally unbraced length for the limit state of inelastic lateral-torsional buckling	in
L_{vi}	interface length considered to be engaged in shear transfer	in/ft
m	base plate cantilever dimension parallel to column web	in
m	slope	–
M	moment	in-lbf
M_1	smaller factored end moment on compression member	in-lbf
M_2	larger factored end member on compression member	in-lbf
M_{arch}	maximum moment due to triangular loading	in-lbf
M_c	factored moment amplified for the effects of member curvature	in-lbf
M_{cx}	available strong-axis flexural strength	in-lbf
M_{lt}	first-order moment caused by lateral translation of the frame only	in-lbf
M_n'	nominal flexural strength neglecting C_b	in-lbf
M_n^-	nominal flexural strength for negative flexure	in-lbf
M_n^+	nominal flexural strength for positive flexure	in-lbf
M_{nb}	nominal flexural strength of beam, including slab where in tension, framing into joint	in-lbf
M_{nc}	nominal flexural strength of column framing into joint	in-lbf
M_{nt}	first-order flexural moment	in-lbf
M_o	total factored static moment	in-lbf
M_p	nominal plastic flexural strength	in-lbf
M_p	plastic bending moment	in-lbf
M_{pr}	probable maximum moment at plastic hinge/center of RBS	in-lbf
M_{px}	strong axis plastic bending moment	in-lbf
M_{Q_E}	moment due to unfactored seismic load	in-lbf
M_r	maximum moment demand	in-lbf
M_r	required design flexural strength/moment	in-lbf
M_u	factored/design moment	in-lbf
n	base plate cantilever dimension parallel to column flange	in
n	modular ratio	–
n	quantity (number of)	–
n'	modified base plate cantilever dimension parallel to column flange	in
N	base plate dimension parallel to column web	in

N	length	in
N	number	–
N_b	number of beams or girders (AASHTO)	–
p	fraction of traffic in a single lane	–
p	load intensity	lbf/in^2
p	penetration depth	in
p	pressure	lbf/in^2
p	snow load	lbf/in^2
p_c	force in compression brace for seismic design	lbf
p_c	plastic force in compression flange	lbf
p_d	maximum intensity of snow drift surcharge load	lbf/in^2
p_e	equivalent uniform static seismic loading for bridges	lbf/in
p_{NW}	net windward pressure	lbf/in^2
p_o	uniform transverse load	lbf/in
p_p	net pressure on parapet	lbf/in^2
p_s	net/adjusted simplified design wind pressure	lbf/in^2
p_s	sloped roof snow load	lbf/in^2
p_{s30}	simplified design wind pressure	lbf/in^2
p_t	length of fastener penetration into wood member	in
P	axial load, strength, or force	lbf
P	plastic force	lbf
P_a	allowable axial compressive force	lbf/in
P_a	allowable compressive force in reinforced masonry	lbf
P_a	required axial strength (ASD) or unfactored force	lbf
P_{AE}	seismic active force	lbf
P_B	base wind pressure	lbf/in^2
P_c	available axial compressive strength	lbf
P_c	compressive force normal to shear plane	lbf
P_c	critical buckling load of compression member	lbf
P_c	expected base strength in compression for seismic design	lbf
$P_{c,buckling}$	expected post-buckling brace strength for seismic design	lbf
P_{drag}	drag-down force on pile foundation	lbf
P_D	design wind pressure	lbf/in^2
$P_{D,WL}$	design wind pressure on vehicles	lbf/in
P_e	prestressing force after all losses	lbf
P_{e1}	elastic critical buckling resistance of column	lbf
P_{end}	end bearing capacity of pile foundation	lbf
P_{lt}	first-order axial load resulting from lateral translation of frame only	lbf
P_L	axial load due to unfactored live load	lbf
P_{ns}	nominal screw shear strength	lbf
P_{nt}	first-order axial compression load	lbf
P_p	available column axial strength	lbf
P_{Q_E}	axial load due to unfactored horizontal seismic forces	lbf
P_r	design axial compression load	lbf
P_r	required axial strength	lbf
P_{rb}	plastic force in bottom layer of longitudinal deck reinforcement	lbf
P_{rt}	plastic force in top layer of longitudinal deck reinforcement	lbf

P_s	adjusted simplified design wind pressure	lbf/in^2
P_s	plastic compressive force in concrete deck	lbf
P_{s30}	simplified wind pressure	lbf/in^2
P_{skin}	skin-friction capacity of pile foundation	lbf
P_{ss}	nominal screw shear strength	lbf
P_S	axial load due to unfactored snow load	lbf
P_t	expected yield strength of tensile brace for seismic design	lbf
P_u	design axial load	lbf
P_u	required axial strength (LRFD) or factored force	lbf
P_{uf}	factored load from tributary roof area	lbf/ft
P_y	nominal axial yield strength of member	lbf
q	bearing/surcharge pressure	lbf/in^2
q	load/unfactored area load	lbf/in,lbf/in^2
q	velocity wind pressure	lbf/in^2
q_{drag}	drag-down pressure	lbf/in^2
q_{end}	end bearing capacity per area	lbf/in^2
q_p	velocity pressure evaluated at top of parapet	lbf/in^2
q_{Q_E}	unfactored out-of-plane seismic area load	lbf/in^2
q_{skin}	skin friction capacity per area	lbf/in^2
q_u	factored area load	lbf/in^2
q_u	factored pressure	lbf/in^2
q_u	strength of fillet weld	lbf/in
Q	flow rate	in^3/sec
Q_E	load effect from horizontal seismic forces	lbf
Q_n	nominal strength of one shear stud connector	lbf
r	radius of gyration	in
R	lateral stiffness of a story	–
R	radius	in
R	reaction	lbf
R	reaction at the roof level	lbf/in
R	relative rigidity	–
R	response modification coefficient/factor	–
RF	effective area reduction factor	–
R_1	roof live load reduction factor	–
R_2	roof live load reduction factor	–
R_a	required connection strength (ASD)	lbf
R_n	nominal connection strength	lbf
R_u	required connection strength (LRFD) or factored reaction	lbf
R_{water}	water flow rate	in^3/sec
R_y	ratio of expected yield stress to specified minimum yield stress	–
s	center-to-center spacing	in
s	spacing (pitch)	in
s_y	transverse pile spacing	in
S	elastic section modulus	in^3
S	skew angle of bridge supports	deg
S	snow load	lbf/in^2
S	transverse spacing of bridge girders	in
S_1	mapped maximum considered earthquake spectral response acceleration at 1 second period	–
S_{D1}	design spectral response acceleration at 1 second period	–

S_{DS}	design spectral response acceleration at short period	–
S_{M1}	maximum considered earthquake spectral response acceleration at 1 second period adjusted for site class effects	–
S_{MS}	maximum considered earthquake spectral response acceleration at short periods adjusted for site class effects	–
S_S	mapped maximum considered earthquake spectral response acceleration at short periods	–
SG	specific gravity	–
t	(nominal) thickness	in
T	fundamental building period	sec
T	maximum/unfactored chord force	lbf
T	temperature	°F
T	tensile/collector force	lbf
T	threaded length of wood screw	in
T_L	mapped long-period transition period	sec
T_m	bridge period of mth mode of vibration	sec
T_0	20% of the ratio of the design spectral response acceleration at 1 sec period to the design spectral response acceleration at short periods	–
T_S	reference period used to define shape of seismic response spectrum	sec
U	shear lag factor	–
v	velocity	mi/hr
v_{DZ}	design wind velocity at elevation Z	mi/hr
v_{si}	shear wave velocity of soil or rock layer i	in/sec
$\overline{v}_s$	average shear wave velocity	in/sec
v	allowable shear	lbf/in
v	shear force per unit length	lbf/in
v_c	stress corresponding to nominal two-way shear strength provided by concrete	lbf/in^2
v_n	nominal shear strength per area	lbf/in^2
v_s	lateral deflection at midspan of bridge	in
v_s	maximum deformation due to unit uniform lateral load	in
v_{si}	shear wave velocity of soil or rock layer i	in/sec
v_u	factored shear force per area	lbf/in^2
v_{ug}	factored shear stress on the slab critical section for two-way action due to gravity loads without moment transfer	lbf/in^2
v_w	diaphragm shear capacity for wind loading	lbf/in
v_w'	adjusted allowable diaphragm shear	lbf/in
V	shear force or shear strength	lbf
V'	required shear force transferred by shear connectors	lbf
V_{ASD}	in-plane shear force in special reinforced masonry wall, amplified for ASD design	lbf
V_b	basic concrete breakout strength in shear of a single anchor	lbf
V_B	base wind velocity	mi/hr
V_c	nominal shear strength provided by concrete	lbf
V_c	punching shear strength of a concrete footing	lbf

V_{cb}	nominal concrete breakout strength in shear of a single anchor	lbf
V_{EQ}	shear due to unfactored earthquake loads	lbf
V_n	nominal shear strength or nominal shear capacity	lbf
V_{ni}	nominal interface shear resistance	lbf/in
V_{nx}	nominal shear strength in principal axis	lbf
V_p	nominal shear strength of an active link	lbf
V_s	nominal shear strength provided by shear reinforcement	lbf
V_u	design/factored shear force	lbf
V_{ui}	factored shear force at the girder-deck interface	lbf/in
w	average weight of bridge	lbf/in
w	design lane load	lbf/in
w	distributed load/weight	lbf/in
w	width	in
$w(x)$	weight of bridge along length	lbf/in
w_D	unfactored distributed dead load	lbf/in
$w_{E\text{-}W}$	diaphragm-level wind load in east-west direction	lbf/in
w_I	portion of effective seismic weight assigned to building level i	lbf
w_I	unfactored ice load	lbf/in
w_L	unfactored distributed live load	lbf/in
w_m	masonry self-weight, per vertical square foot	lbf/in^2
$w_{N\text{-}S}$	diaphragm-level wind load in north-south direction	lbf/in
w_u	factored area or distributed load	lbf/in, lbf/in^2
W	load due to wind pressure	lbf
W	reference withdrawal design value	lbf/in
W	vertical/lateral load or force	lbf/in
W	weight	lbf
W_p	weight of wall tributary to seismic anchor	lbf
W_u	factored self-weight	lbf
W_v	vertical load	lbf/in
x	approximate building period parameter	–
x	distance	in
x_n	distance from origin to lateral load resisting element n	in
X	base plate parameter	–
X	dimension	in
y_b	distance from centroid to bottom fiber	in
y_n	distance from origin to lateral load resisting element n	in
y_t	distance from centroid to top fiber	
$\bar{y}$	distance to centroid	in
Y	dimension	in
Y	dimension of truncated circle of projected tension area	in
$\bar{Y}$	distance from top of concrete deck to plastic neutral axis	in
$Y1$	distance from top flange of composite steel beam to plastic neutral axis	in
$Y2$	distance from concrete flange force to steel beam top flange	in
Y_{con}	distance from top of steel beam to top of concrete	in
z	height	in
Z	dimension	in

Z	plastic section modulus	in^3
Z	reference lateral design value for a single fastener in a wood connection	lbf
Z'	adjusted lateral design value for a single fastener in a wood connection	lbf
Z_e	effective plastic modulus at the location of a plastic hinge	in^3
$Z_{s\perp}$	reference lateral design value for a single fastener in a wood connection with main member loaded parallel to grain and side member loaded perpendicular to grain	lbf

Symbols

α	angle	rad
α	thermal coefficient of expansion	1/°F
α_c	coefficient defining relative contribution of concrete strength to nominal shear wall strength	–
α_f	ratio of flexural stiffness of beam section to flexural stiffness of slab	–
α_s	constant used to compute shear strength of slabs and footings	–
β	ratio of long to short dimension of column	–
β_t	ratio of torsional stiffness of edge beam to flexural stiffness of slab	–
γ	reinforcement size factor for masonry	–
γ	unit weight (density)	lbf/in^3
γ_f	factor to determine unbalanced moment transferred by flexure at slab-column connections	–
γ_g	factor to determine allowable shear stress of reinforced masonry	–
γ_v	factor to determine unbalanced moment transferred by eccentricity of shear at slab-column connections	–
γ_{LL}	load factor for live load	–
γ_{TU}	load factor for uniform temperature loads	
δ	angle	rad
δ	moment magnification factor to reflect effects of member curvature between ends of compression members	–
δ_{xe}	deflection at center of mass of building level x determined by an elastic analysis	in
$\delta_{x+1,e}$	deflection at center of mass one story above level x determined by elastic analysis	in
Δ	deflection	in
Δ	design story drift	in
Δ_a	allowable story drift	in
Δ_I	immediate or initial deflection	in
Δ_L	live load deflection	in
Δ_{ph}	horizontal wall pressure resulting from surcharge load	lbf/in^2
Δ_T	total deflection including long-term loading effects for wood beams	in
Δ_x	story drift	in
Δ_{x+1}	story drift of story between two floors (x is lower floor)	in
$(\Delta F)_n$	nominal fatigue resistance	lbf/in^2
θ	angle	deg
λ	adjustment factor	–
λ	base plate parameter	in

λ	modification factor related to concrete density	–
λ	wind adjustment factor for building height and exposure	–
λ_Δ	multiplier for additional concrete deflection due to long-term effects	–
μ	coefficient of friction	–
ν	Poisson's ratio	–
ξ	time-dependent factor for sustained loads	–
ρ	redundancy factor	–
$\rho_,$	reinforcement ratio	–
ρ'	compression reinforcement ratio	–
ρ_t	reinforcement ratio at which section is no longer tensile controlled	–
ϕ	resistance factor or strength reduction factor	–
$\Psi_{c,v}$	modification factor for shear strength of anchors based on presence of cracks	–
Ψ_e	modification factor for development length based on rebar coating	–
$\Psi_{cd,v}$	modification factor for shear strength of anchors based on edge distance	–
$\Psi_{h,v}$	modification factor for shear strength of anchors based on thickness of concrete	–
Ψ_s	modification factor for development length based on rebar size	–
Ψ_t	modification factor for development length based on rebar location	–
ω	tension reinforcement index	–
Ω	safety factor	–

p	cell plate
prov	cell provided
r	cell reduced roof, required, or right
req	cell required
res	cell resistance
s	cell shear, slab, sloped roof, side member, soil, spiral, steel, story, or structural steel shape
st	cell steel reinforcement
ST	cell short term
strap	cell footing strap beam
sub	cell substructure
super	cell superstructure
sur	cell surcharge
t	cell tensile, tensile flange, total, transverse, or tributary
T	cell torsional
tot	cell total
trans	cell transverse
u	cell unsupported
v	cell shear
w	cell wall, web, or weld
WT	cell wide flange tee reinforcing

Subscripts

a	cell approximate
adj	cell adjacent
b	cell bearing or bending
B	cell base
bot	cell bottom
c	cell clear, column, column strip, compression, compression flange, concrete, or cylinder
comp	cell composite section
CG	cell center of gravity
CM	cell center of mass
CR	cell center of rigidity
d	cell design ice or snow drift
D	cell dead load
dia	cell diaphragm reaction
e	cell effective net, elastic, end spans, or existing beams
f	cell flange, flat roof, or floor
fric	cell friction
g	cell girder, gross, or ground
h	cell horizontal
i	cell ice or interior spans
int	cell interior
long	cell longitudinal
l	cell left or longitudinal
L	cell live load
LB	cell lower bound
LT	cell long term
m	cell main member, masonry, masonry veneer, or moment
max	cell maximum
min	cell minimum
n	cell net or nominal
opp	cell opposite

Vertical Forces Component: Breadth Module Instructions

In accordance with the rules established by your state, you may use textbooks, handbooks, bound reference materials, and any approved battery- or solar-powered, silent calculator to work this examination. However, no blank papers, writing tablets, unbound scratch paper, or loose notes are permitted. Sufficient room for scratch work is provided in the Examination Booklet.

You are not permitted to share or exchange materials with other examinees.

You will have four hours in which to work this module of the examination. Your score will be determined by the number of problems that you answer correctly. There is a total of 40 problems. There are no optional problems. Each problem is worth 1 point, and there is no penalty for incorrect answers. The maximum possible score for this section of the examination is 40 points.

Partial credit is not available. No credit will be given for methodology, assumptions, or work written in your Examination Booklet.

Record all of your answers on the Answer Sheet. No credit will be given for answers marked in the Examination Booklet. Mark your answers with a no. 2 pencil. Answers marked in pen may not be graded correctly. Marks must be dark and must completely fill the bubbles. Record only one answer per problem. If you mark more than one answer, you will not receive credit for the problem. If you change an answer, be sure the old bubble is erased completely; incomplete erasures may be misinterpreted as answers.

If you finish early, check your work and make sure that you have followed all instructions. After checking your answers, you may turn in your Examination Booklet and Answer Sheet and leave the examination room. Once you leave, you will not be permitted to return to work or change your answers.

When permission has been given by your proctor, break the seal on the Examination Booklet. Check that all pages are present and legible. If any part of your Examination Booklet is missing, your proctor will issue you a new Booklet.

Note that this practice exam is not an exact representation of the actual exam. For the exam, you must include your name, date of birth, examinee number, test site, and other information on a Scantron sheet. Each question will also be on a separate page with ample blank space for scratch work.

WAIT FOR PERMISSION TO BEGIN

Name: _____
       Last   First   Middle Initial

Examinee number: _____

Examination Booklet number: _____

Structural Engineering (SE) Examination

Vertical Forces Component Breadth Module

Vertical Forces Component: Breadth Module Answer Sheet

1. (A) (B) (C) (D)   11. (A) (B) (C) (D)   21. (A) (B) (C) (D)   31. (A) (B) (C) (D)

2. (A) (B) (C) (D)   12. (A) (B) (C) (D)   22. (A) (B) (C) (D)   32. (A) (B) (C) (D)

3. (A) (B) (C) (D)   13. (A) (B) (C) (D)   23. (A) (B) (C) (D)   33. (A) (B) (C) (D)

4. (A) (B) (C) (D)   14. (A) (B) (C) (D)   24. (A) (B) (C) (D)   34. (A) (B) (C) (D)

5. (A) (B) (C) (D)   15. (A) (B) (C) (D)   25. (A) (B) (C) (D)   35. (A) (B) (C) (D)

6. (A) (B) (C) (D)   16. (A) (B) (C) (D)   26. (A) (B) (C) (D)   36. (A) (B) (C) (D)

7. (A) (B) (C) (D)   17. (A) (B) (C) (D)   27. (A) (B) (C) (D)   37. (A) (B) (C) (D)

8. (A) (B) (C) (D)   18. (A) (B) (C) (D)   28. (A) (B) (C) (D)   38. (A) (B) (C) (D)

9. (A) (B) (C) (D)   19. (A) (B) (C) (D)   29. (A) (B) (C) (D)   39. (A) (B) (C) (D)

10. (A) (B) (C) (D)   20. (A) (B) (C) (D)   30. (A) (B) (C) (D)   40. (A) (B) (C) (D)

Vertical Forces Component: Breadth Module Exam

1. The elevation of a building where the roof must be designed for drifting snow loads is shown.

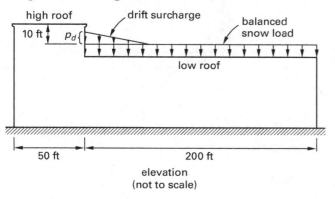

elevation
(not to scale)

Design Code

- ASCE/SEI7

Design Criteria

- ground snow load, $p_g = 60 \text{ lbf/ft}^2$

- do not include the balanced snow load as part of the drift surcharge load

The maximum intensity of the drift surcharge load, p_d, is most nearly

- (A) 43 lbf/ft^2
- (B) 67 lbf/ft^2
- (C) 94 lbf/ft^2
- (D) 130 lbf/ft^2

2. A simply supported, single-span, concrete slab bridge is shown. The main flexural reinforcement is parallel to traffic.

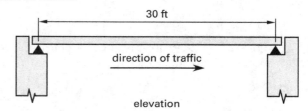

elevation

Design Code

- AASHTO

Considering an HL-93 vehicular loading, the maximum factored design live load moment per traffic lane, including impact, for the strength I limit state is most nearly

- (A) 440 ft-kips
- (B) 700 ft-kips
- (C) 760 ft-kips
- (D) 890 ft-kips

3. A single-span, concrete bridge is shown. The abutments are rigid, and displacements at the ends of the slab are restrained in the longitudinal direction by the abutments.

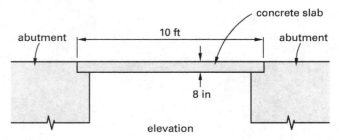

elevation

Design Code

- AASHTO

Design Criteria

- concrete density, $\gamma_c = 0.145 \text{ kips/ft}^3$

- concrete strength, $f_c' = 3 \text{ kips/in}^2$

- temperature range (uniform through concrete slab thickness), $\Delta T = 30°F$

- service II limit state, $\gamma_{TU} = 1.00$

The maximum horizontal reaction at the abutments per foot of slab width due to a temperature increase of 30°F in the slab is most nearly

(A)   5.5 kips/ft

(B)   33 kips/ft

(C)   55 kips/ft

(D)   65 kips/ft

4. An exterior W10 × 12 steel beam at a police station is 20 ft above the ground.

Design Code

- ASCE/SEI7

Design Criteria

- ice density, $\gamma_i = 57.2$ lbf/ft^3

- mapped nominal ice thickness at height of 33 ft, $t = 0.75$ in

- topographical factor, $K_{zt} = 1.0$

- the police station is an essential facility

The unfactored atmospheric ice load on the beam is most nearly

(A)   22 lbf/ft

(B)   28 lbf/ft

(C)   34 lbf/ft

(D)   69 lbf/ft

5. An infinitely long line load is applied to a strip of ground parallel to a wall, as shown.

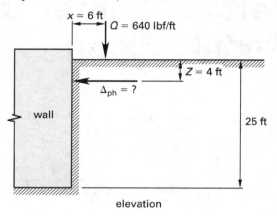

elevation

Design Code

- AASHTO

Design Criteria

- do not consider any load factors

- assume the wall is restrained from movement

At a point 4 ft below the top-of-soil elevation, the horizontal pressure on the wall, Δ_{ph}, is most nearly

(A)   7.2 lbf/ft^2

(B)   11 lbf/ft^2

(C)   16 lbf/ft^2

(D)   43 lbf/ft^2

6. The elevation of a two-span bridge with a central concrete pier is shown. Water approaches the central pier at a 20° angle, as shown.

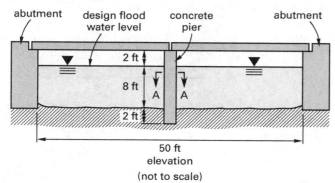

elevation
(not to scale)

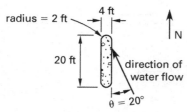

section A-A concrete pier

Design Code

- AASHTO

Design Criteria

- pier is fixed at the base and free at the top (cantilever condition)

- water flow rate at design flood, $Q = 6000$ ft³/sec

The unfactored moment at the base of the concrete pier due to lateral stream water pressure (east-west direction) is most nearly

(A)   100 ft-kips

(B)   130 ft-kips

(C)   160 ft-kips

(D)   190 ft-kips

7. A continuous steel angle that supports a masonry veneer is shown.

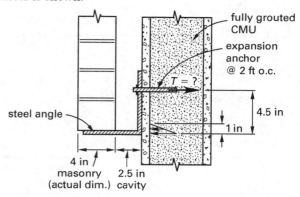

Design Criteria

- height of supported masonry veneer above the angle, $h = 40$ ft

- masonry veneer density, $\gamma_m = 120$ lbf/ft³

- assume the steel angle is rigid

- assume the depth of the triangular compression block is 1 in

The tensile force, T, in one expansion anchor due to the self-weight of the masonry is most nearly

(A)   0 kips

(B)   1.7 kips

(C)   3.2 kips

(D)   3.5 kips

8. A distributed load acts on a beam, as shown.

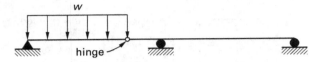

The moment diagram is most nearly

(A)

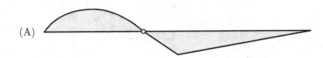

(B)

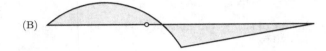

(C)

(D)

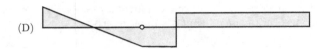

9. A piece of mechanical equipment is to be hung from a wood beam. Four possible connections are detailed as shown. Of these, the most preferable connection detail for preventing horizontal splits in the wood is

(A)

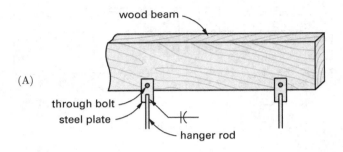

(B)

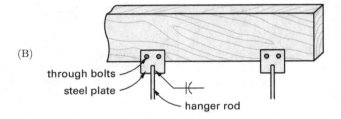

(C)

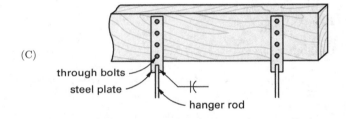

(D)

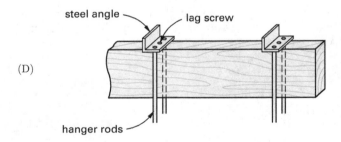

10. The cross section and elevation of a concrete bridge deck with four composite steel girders are shown.

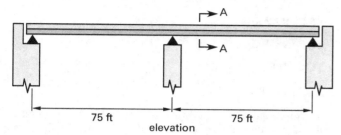

75 ft   75 ft

elevation

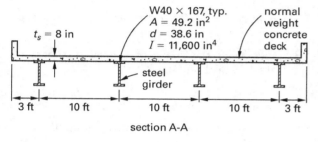

section A-A

Design Code

• AASHTO

Design Criteria

• modular ratio, $n = E_{beam}/E_{deck} = 8$

For the case of two or more design lanes loaded, the live load distribution factor for negative moment in an interior beam is most nearly

(A) 0.52 lanes

(B) 0.62 lanes

(C) 0.74 lanes

(D) 0.84 lanes

11. A structural analysis of the steel beams shown is performed using a computer analysis program. All connections and supports are modeled as pinned, and each beam is bent about its strong axis.

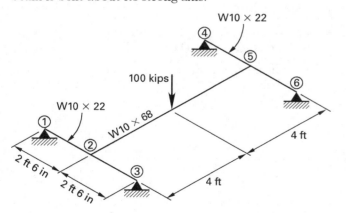

Under the unfactored load of 100 kips, the computer-generated deflection at node 2 is 0.096 in. To verify the results, the following calculation is performed by hand.

$$\Delta = \frac{PL^3}{48EI} = \frac{\left(\dfrac{100 \text{ kips}}{2}\right)\left((2.5 \text{ ft} + 2.5 \text{ ft})\left(12 \, \dfrac{\text{in}}{\text{ft}}\right)\right)^3}{(48)\left(29{,}000 \, \dfrac{\text{kips}}{\text{in}^2}\right)(118 \text{ in}^4)}$$

$$= 0.066 \text{ in}$$

The discrepancy between the two deflections is because the calculation done by hand ignores deflection due to

(A) creep under sustained, long-term loading

(B) P-delta effects

(C) self-weight of the beams

(D) shear deformations

12. A two-span, continuous beam subjected to two point loads is shown.

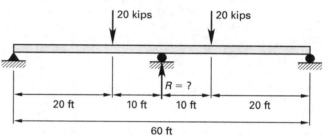

20 ft   10 ft   10 ft   20 ft

60 ft

Ignoring the self-weight of the beam, the reaction, R, at the interior support is most nearly

(A) 24 kips

(B) 27 kips

(C) 34 kips

(D) 38 kips

13. A planar truss is shown. All connections are pinned.

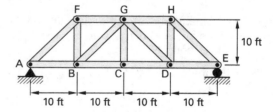

10 ft   10 ft   10 ft   10 ft

Tension (–) is shown on the top side of the influence line diagrams, and compression (+) is shown on the bottom side of the influence line diagrams. For a downward vertical loading applied to the bottom chord, which option shows the influence line for the axial force in member CG?

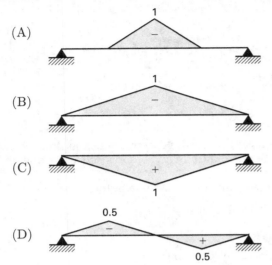

(A)

(B)

(C)

(D)

14. A single-story steel building with a basement is shown. The roof is an ordinary flat roof and is not used as a promenade or roof garden, or for assembly purposes. The first floor is used as a light storage warehouse. The basement slab is supported on soil.

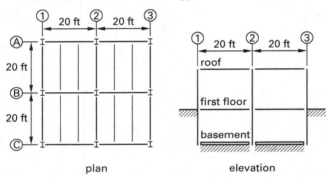

plan

elevation

Design Code

- ASCE/SEI7

Design Criteria

- dead loads (including all building material weights and allowance for column self-weight), roof: 40 lbf/ft^2, first floor: 60 lbf/ft^2

- use load combination $1.2D + 1.6L_r + L$ (ASCE/SEI7 Sec. 2.3.2, Eq. 3)

Considering live load reduction, the design axial load at the base of column B-2 is most nearly

(A) 90.2 kips

(B) 98.4 kips

(C) 108 kips

(D) 111 kips

15. Susceptibility to vibration serviceability problems will be increased due to all of the following changes EXCEPT

(A) changing steel-framed floors to flat plate concrete slabs

(B) increasing the spans of typical members of a steel-framed building

(C) decreasing the depths of typical members of a steel-framed building

(D) changing the use of a steel-framed building from an office to an aerobics gym

16. A bridge has simply supported, longitudinal steel girders. As shown, the girders have transverse stiffener plates where the stiffener-to-flange fillet weld connection must be evaluated for fatigue.

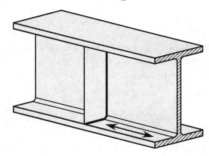

Design Code

- AASHTO

Design Criteria

- girder span, 60 ft

- number of trucks per day in one direction, averaged over the design life, ADTT = 1000 trucks

- number of lanes available to trucks in each direction, three lanes

The bridge has a finite life of 75 years. Using the fatigue II load combination, the nominal fatigue resistance, $(\Delta F)_n$, for the welded stiffener connection is most nearly

(A)   5.5 kips/in^2

(B)   5.9 kips/in^2

(C)   7.4 kips/in^2

(D)   12 kips/in^2

17. A column base plate for a W14 × 109 column is shown.

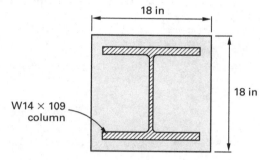

Design Code

- AISC *Steel Construction Manual* (including AISC 360)

Design Criteria

- base plate steel, $F_y = 36$ kips/in^2
- column steel, $F_y = 50$ kips/in^2
- column effective length, $K_yL = K_xL = 15$ ft
- assume $\lambda = 1.0$
- design base plate for full available axial strength of the column
- entire base plate area bears on concrete; concrete bearing strength does not control
- use ASD or LRFD

The minimum required thickness of the base plate is most nearly

(A)   ½ in

(B)   1 in

(C)   1½ in

(D)   1¾ in

18. An existing steel office building is being converted into a library, and the steel floor framing must be reinforced to support higher live loads. The design calls for welding WT8 × 25s (A992 steel) to the underside of existing W12 wide flange beams, as shown. Assume that the welds are adequate to transfer the shear at the interface.

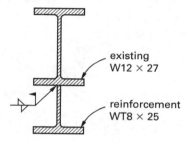

Design Code

- AISC *Steel Construction Manual*

Existing W12 × 27 Beam Properties

Material properties of A7-61T steel

- $F_y = 33$ kips/in^2
- $E = 29{,}000$ kips/in^2

Section Properties

- $A = 7.97$ in^2
- $d = 11.96$ in
- $S_x = 34.1$ in^3
- $I_x = 204.1$ in^4

At the time the reinforcement is installed, the existing beam is subjected to a moment of 20 ft-kips. After the reinforcement is installed, the composite section is subjected to an additional moment of 80 ft-kips. Under the final loaded condition, the maximum bending stress at the top of the W12 × 27 beam is most nearly

(A)   14 kips/in^2

(B)   18 kips/in^2

(C)   21 kips/in^2

(D)   33 kips/in^2

19. A simply supported steel beam supports a concentrated load, as shown.

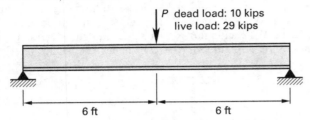

Design Code

- AISC *Steel Construction Manual* (including AISC 360)

Design Criteria

- ASTM A992 steel, grade 50; $F_y = 50$ kips/in^2
- beam braced laterally only at the supports ($L_b = 12$ ft)
- do NOT assume $C_b = 1.0$
- ignore any deflection criteria
- use ASD or LRFD

The lightest wide-flange steel shape that will be adequate for the loads is most nearly

- (A) W14 × 30
- (B) W14 × 34
- (C) W18 × 35
- (D) W16 × 36

20. A piece of a WT9 × 43 is used in a connection detail, as shown. The full eccentricity is taken by the three-sided fillet weld alone.

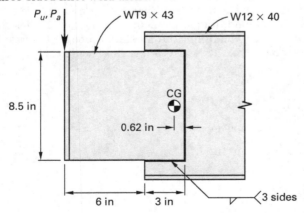

Design Code

- AISC *Steel Construction Manual* (including AISC 360)

Design Criteria

- loads, ASD: $P_a = 16$ kips; LRFD: $P_u = 24$ kips
- E70XX electrodes
- use ASD or LRFD

Using the instantaneous center of rotation method, the minimum adequate size for the three-sided fillet weld is most nearly

- (A) $^3\!/_{16}$ in
- (B) $^1\!/_4$ in
- (C) $^5\!/_{16}$ in
- (D) $^3\!/_8$ in

21. The cross section of a simply supported composite bridge superstructure is shown.

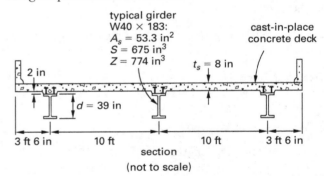

section
(not to scale)

Design Code

- AASHTO

Design Criteria

- concrete strength, $f'_c = 4$ kips/in^2
- steel strength, $F_y = 50$ kips/in^2
- distance from top of concrete deck to plastic neutral axis, $\overline{Y} = 6.53$ in
- ignore contribution from slab reinforcement
- steel section is compact

The nominal flexural resistance of the composite interior girder is most nearly

(A) 3230 ft-kips

(B) 5100 ft-kips

(C) 5690 ft-kips

(D) 5830 ft-kips

22. A cold-formed joist that bears on a concrete wall is shown.

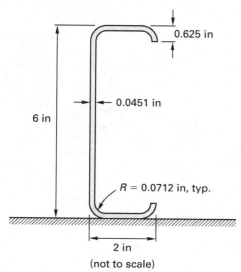

0.625 in

0.0451 in

6 in

$R = 0.0712$ in, typ.

2 in

(not to scale)

Design Code

- *AISI Manual* and *AISI Specification*

Design Criteria

- bearing length, $N = 2$ in

- flanges are stiffened

- joists fastened to the support

- SSMA stud designation 600S200-43, $F_y = 33$ kips/in^2

- angle between plane of web and plane of bearing surface, $\theta = 90°$

For a one-flange reaction at an end support, the nominal web crippling strength of the joist is most nearly

(A) 330 lbf

(B) 370 lbf

(C) 570 lbf

(D) 860 lbf

23. A portion of the cross section of a normal weight cast-in-place concrete bridge deck with precast concrete bulb-tee girders is shown.

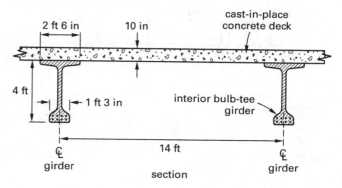

2 ft 6 in   10 in   cast-in-place concrete deck

4 ft

1 ft 3 in   interior bulb-tee girder

$\underline{\mathbb{C}}$ girder   14 ft $\underline{\mathbb{C}}$ girder

section

Design Code

- AASHTO

Design Criteria

- include dynamic load allowance and multi-presence factors

At the design section for negative moment at an interior girder, the maximum live load slab moment is most nearly

(A) 9.3 ft-kips/ft

(B) 9.7 ft-kips/ft

(C) 11 ft-kips/ft

(D) 12 ft-kips/ft

24. According to AASHTO, the thickness of a concrete bridge deck, unless it is approved otherwise by the owner, should be at least

(A) 7 in

(B) 8 in

(C) $1/_{30}$ of the design span

(D) $1/_{20}$ of the design span

25. A simple-span, lightweight, concrete slab is shown. The slab is reinforced with no. 5 bars top and bottom at 12 in o.c.

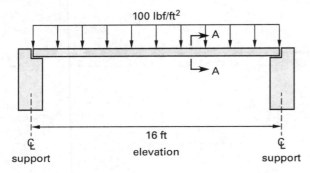

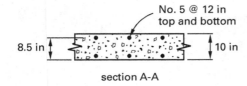

section A-A

Design Code

- ACI 318

Design Criteria

- sustained service-level load (more than five years), 100 lbf/ft^2 (includes self-weight of slab)

- concrete density, $\gamma_c = 110$ lbf/ft^3

- concrete strength, $f_c' = 3500$ lbf/in^2

- effective moment of inertia per foot width of slab, $I_e = 600$ in^4/ft

Including long-term contributions from creep and shrinkage, the total midspan deflection due to the sustained load is most nearly

(A)   0.13 in

(B)   0.19 in

(C)   0.22 in

(D)   0.30 in

26. A deep beam supports a factored column load of 400 kips, as shown. A strut-and-tie model is used to analyze the beam.

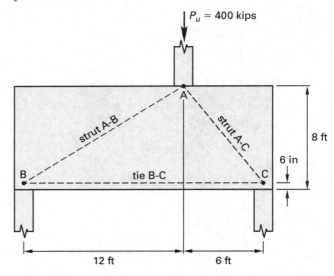

Design Code

- ACI 318

Design Criteria

- concrete, $f_c' = 5000$ lbf/in^2

- steel reinforcement, $f_y = 60$ kips/in^2

- beam width, $b_w = 12$ in

- factored column load includes allowance for beam self-weight

The required area of flexural steel in tie B-C is most nearly

(A)   3.8 in^2

(B)   4.3 in^2

(C)   4.8 in^2

(D)   5.1 in^2

27. A continuous concrete beam subjected to a factored load of 5.4 kips/ft (including beam self-weight) is shown.

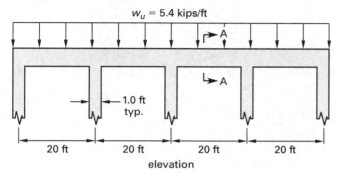

elevation

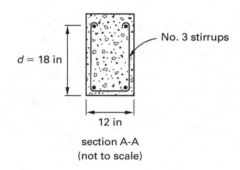

section A-A
(not to scale)

Design Code

- ACI 318

Design Criteria

- normal weight concrete, $f'_c = 4000 \text{ lbf/in}^2$

- steel reinforcement, $f_{yt} = 60 \text{ kips/in}^2$

- stirrups not inclined

- use approximate method of analysis (per ACI 318 Sec. 6.5.4) to calculate shear demand

At the critical section for shear at the first interior support of the end bay, the minimum adequate spacing for the no. 3 shear stirrups is most nearly

- (A)   4 in
- (B)   6 in
- (C)   9 in
- (D)   12 in

28. A portion of the cross section of a normal weight cast-in-place concrete bridge is shown. At the time that the slab is placed, the interface at the top of each girder is clean and roughened to an amplitude of 0.25 in.

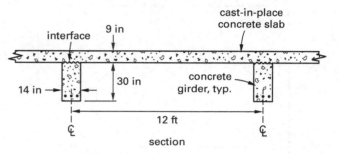

section

Design Code

- AASHTO

Design Criteria

- compressive strength of concrete, $f'_c = 4 \text{ kips/in}^2$

- reinforcement strength, $f_y = 60 \text{ kips/in}^2$

- factored shear force at the girder-deck interface, $V_{ui} = 100 \text{ kips/ft}$

- permanent compressive force normal to shear plane, $P_c = 25 \text{ kips/ft}$

- assume strength I limit state

The required area of interface shear reinforcement at the girder-deck interface, A_{vf}, per foot length of girder is most nearly

- (A) $0.47 \text{ in}^2/\text{ft}$
- (B) $0.65 \text{ in}^2/\text{ft}$
- (C) $0.76 \text{ in}^2/\text{ft}$
- (D) $3.8 \text{ in}^2/\text{ft}$

29. A section through a simply supported, prestressed concrete box beam that spans 65 ft is shown.

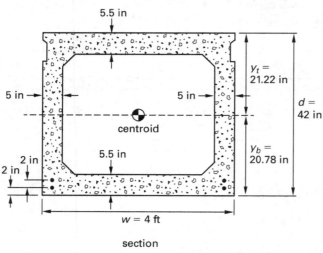

section

Design Criteria

- concrete density, $\gamma_c = 0.145$ kips/ft^3
- girder area, $A = 843$ in^2
- girder moment of inertia, $I = 203{,}100$ in^4
- prestressing force after all losses, $P_e = 550$ kips

Considering the self-weight of the beam and the prestressing force, the stress at the top of the beam is most nearly

(A) -0.93 kips/in^2 (tension)

(B) -0.46 kips/in^2 (tension)

(C) 0.19 kips/in^2 (compression)

(D) 1.2 kips/in^2 (compression)

30. A Douglas fir $3\frac{1}{2}$ in $\times$ $7\frac{1}{2}$ in glued laminated timber is bent about its strong axis (x-x axis) and is subjected to the distributed loads shown.

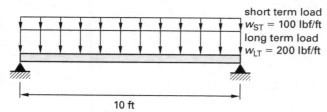

Design Code

- NDS (including NDS Supp.)

Design Criteria

- $C_M = C_t = 1.0$
- glulam combination symbol, 16F-E3
- long-term load includes the beam self-weight

Including long-term loading effects, the total beam deflection is most nearly

(A)   0.34 in

(B)   0.40 in

(C)   0.46 in

(D)   0.88 in

31. A Douglas fir-south select structural 6×8 timber is subjected to a compressive load. The post has different effective lengths for buckling about each axis, as shown.

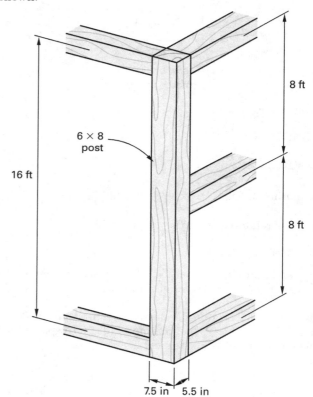

Design Code

- NDS

Design Criteria

- $C_D = C_M = C_t = C_F = C_i = C_T = 1.0$
- $E_{\min} = 440{,}000$ lbf/in^2
- $E = 1{,}200{,}000$ lbf/in^2

- $F_{c\perp} = 520 \text{ lbf/in}^2$

- $F_c = 1050 \text{ lbf/in}^2$

- assume $K_e = 1.0$

The column stability factor, C_P, is most nearly

(A) 0.31

(B) 0.45

(C) 0.66

(D) 0.98

32. A $\frac{1}{2}$ in diameter, 2 in long standard hex lag screw is used to fasten a $\frac{1}{4}$ in, ASTM A36 steel plate into the side grain of a 3×8 piece of Douglas fir-larch (north), no. 2 sawn lumber, as shown.

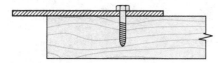

Design Code

- NDS

The nominal withdrawal design value for the lag screw is most nearly

(A) 370 lbf

(B) 440 lbf

(C) 510 lbf

(D) 700 lbf

33. A section through a CMU wall where reinforcing bars are lap spliced is shown.

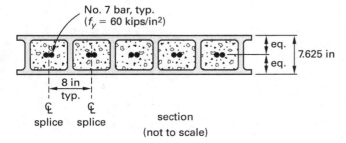

Design Code

- TMS 402/TMS 602

Design Criteria

- mortar, type S

- net area compressive strength of concrete masonry units, 3250 lbf/in^2

- no transverse reinforcement

The minimum required lap length for the tension splice of the no. 7 bars is most nearly

(A) 33 in

(B) 38 in

(C) 41 in

(D) 46 in

34. A CMU lintel beam is shown. In addition to its own self-weight, the beam also supports a triangular load from the weight of the clay masonry above considering arching action.

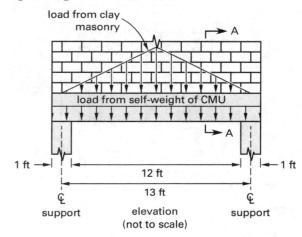

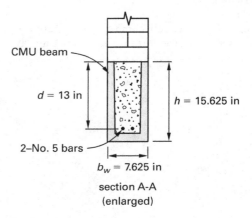

Design Code

- TMS 402

Design Criteria

- CMU and clay masonry self-weight (per surface area), $w_m = 80 \text{ lbf/ft}^2$

- masonry strength, $f'_m = 2000 \text{ lbf/in}^2$

- the CMU beam is simply supported and not built integrally with the supports

- use ASD

The maximum compression stress in the CMU beam considering the self-weight of the CMU beam plus the dead load from the arching clay masonry is most nearly

(A) 280 lbf/in^2

(B) 560 lbf/in^2

(C) 670 lbf/in^2

(D) 1100 lbf/in^2

35. A reinforced CMU column is shown.

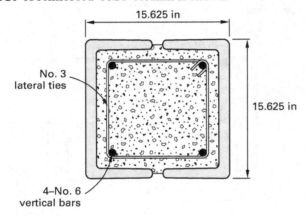

Design Code

- TMS 402

Ignoring seismic requirements, the minimum required vertical spacing of the lateral ties is most nearly

(A) 12 in

(B) 14 in

(C) 16 in

(D) 18 in

36. An 18 in diameter auger cast pile is installed through soft soils and into firm soils, as shown. The data from the geotechnical report is listed.

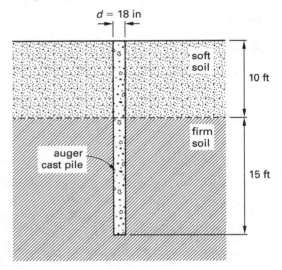

Design Criteria

- piles can be designed for 1000 lbf/ft^2 in skin friction in firm soil

- piles can be designed for 10,000 lbf/ft^2 in end bearing

- soft soils cause a drag-down force of 400 lbf/ft^2

- design values in the geotechnical report include all required safety factors

The allowable axial pile load is most nearly

(A) 18 kips

(B) 37 kips

(C) 70 kips

(D) 110 kips

37. The plan and section of a strap footing subjected to dead and live loads is shown.

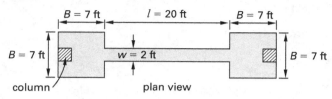

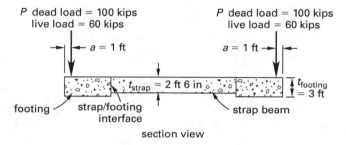

Design Criteria

- concrete density, $\gamma_c = 150$ lbf/ft^3

- use load combination $1.2D + 1.6L$

The design moment at the strap/footing interface is most nearly

- (A)   480 ft-kips

- (B)   510 ft-kips

- (C)   540 ft-kips

- (D)   1300 ft-kips

38. A CMU basement wall that spans vertically between the first floor wood framing and the concrete slab on grade is shown. The top of the wall can be considered restrained against movement.

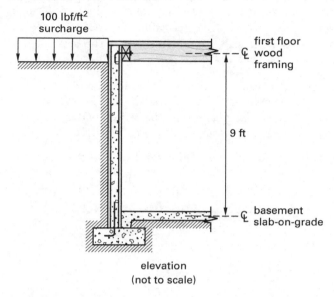

elevation
(not to scale)

Design Criteria

- soil density, $\gamma_s = 110$ lbf/ft^3

- friction coefficient between concrete and soil, $\mu = 0.28$

- coefficient of active soil pressure, $K_a = 0.33$

- coefficient of at-rest soil pressure, $K_o = 0.50$

- coefficient of passive soil pressure, $K_p = 3.0$

- no hydrostatic pressure

- do not use any load factors or safety factors

Considering the soil and surcharge loads, the horizontal reaction at the basement slab is most nearly

- (A)   970 lbf

- (B)   1100 lbf

- (C)   1600 lbf

- (D)   1700 lbf

39. A cast-in-place slab was designed using an average concrete compressive strength of $f'_c = 4000$ lbf/in^2. After the slab was placed, the concrete test cylinder report indicates that the standard laboratory-cured cylinders have a 28-day average compressive strength of only $f'_c = 3400$ lbf/in^2. According to ACI 318, which action is permissible for investigating the slab's load-carrying capacity?

(A) performing calculations to determine if the slab meets the code strength and serviceability requirements using the lower value of $f'_c = 3400$ lbf/in^2

(B) performing tests of cores drilled in the questionable area where low-strength concrete is confirmed in accordance with ASTM C42

(C) performing a load test in accordance with ACI 318 Sec. 27.4

(D) all of the above

40. An office building with concrete spread footings is under construction. The footings will bear on compacted fill.

Design Code

- IBC

Which of the following verification and inspection tasks require continuous inspection as part of the special inspections of soils?

(A) Prior to placement of compacted fill, observe subgrade and verify that the site has been prepared properly.

(B) Perform classification and testing of compacted fill materials.

(C) Verify the use of proper materials, densities, and lift thicknesses during the placement and compaction of compacted fill.

(D) all of the above

STOP!

DO NOT CONTINUE!

This concludes the Vertical Forces Component: Breadth Module of the examination. If you finish early, check your work and make sure that you have followed all instructions. After checking your answers, you may turn in your Examination Booklet and Answer Sheet and leave the examination room. Once you leave, you will not be permitted to return to work or change your answers.

Vertical Forces Component: Buildings Depth Module Instructions

In accordance with the rules established by your state, you may use textbooks, handbooks, bound reference materials, and any approved battery- or solar-powered, silent calculator to work this examination. However, no blank papers, writing tablets, unbound scratch paper, or loose notes are permitted. Sufficient room for showing your work is provided in the Solution Booklet.

You are not permitted to share or exchange materials with other examinees.

You will have four hours in which to work this module of the examination. Your solutions to the essay problems will be evaluated by two subject matter experts for overall compliance with established scoring criteria and general quality. A third subject matter expert will be used when necessary. There is a total of four problems. All four problems are equally weighted. They must be worked correctly in order to receive full credit on the exam. There are no optional problems.

Along with your Examination Booklet, a Solution Booklet of graph paper is provided for each problem. Record your solutions in the appropriate booklet for each problem. No credit will be given for solutions written in the incorrect Booklet.

If you finish early, check your work and make sure that you have followed all instructions. After checking your answers, you may turn in your Examination and Solution Booklets and leave the examination room. Once you leave, you will not be permitted to return to work or change your answers.

When permission has been given by your proctor, break the seals on the Examination and Solution Booklets. Check that all pages are present and legible. If any part of your Examination or Solution Booklet is missing, your proctor will issue you a new Booklet.

Note that this practice exam is not an exact representation of the actual exam. For the exam, you must include your name, date of birth, examinee number, test site, and other information on a Scantron sheet. You will be provided four separate booklets of blank graph paper (one for each problem) in which to show your work. You must write your name and examinee number on each booklet.

Structural Engineering (SE) Examination

Vertical Forces Component Buildings Depth Module

Vertical Forces Component: Buildings Depth Module Exam

41. An intermediate floor of a normal weight concrete building with a two-way, flat plate slab (no drop panels or edge beams) is shown.

Design Codes

- IBC
- ACI 318

Design Criteria

- slab thickness, $t_s = 9$ in
- average effective depth of slab, $d = 7.5$ in
- columns, 12 in, square
- floor dead load, $q_{dead} = 140$ lbf/ft² (includes self-weight of slab)
- floor live load, $q_{live} = 100$ lbf/ft² (non-reducible)
- concrete strength, $f'_c = 4000$ lbf/in²
- reinforcing strength, $f_y = 60{,}000$ lbf/in²
- lateral loads resisted by shear walls (not shown)

41. (a) Using the direct design method, determine the factored positive and negative moments for the column strip on gridline B.

41. (b) Draw the moment diagram for the column strip from part (a).

41. (c) Design the slab flexural reinforcement for the column strip end span at gridline B. Draw the sizes and extents of the bars in an elevation view.

41. (d) Check the adequacy of the slab for punching shear at column B-1. Calculate the direct shear component based on tributary area, and include the shear contribution from the unbalanced moment.

41. (e) List two ways that the design can be modified to address inadequate punching shear strength in a flat plate slab.

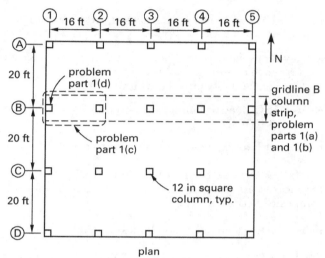

plan

42. The plan of a two-story, steel building is shown in illustration I. The building has moment frames in the east-west direction on column lines 1 and 2 and braced frames in the north-south direction on column lines A and D. Illustration II and illustration III show unit load diagrams.

Design Codes

- IBC
- AISC *Steel Construction Manual* (including AISC 360)

Design Criteria

- all columns and beams are ASTM A992, grade 50; $F_u = 65$ kips/in²; $F_y = 50$ kips/in²
- steel plate is ASTM A36; $F_u = 58$ kips/in²; $F_y = 36$ kips/in²
- slab construction, 3 in normal weight concrete over 3 in metal deck
- concrete strength, $f'_c = 4$ kips/in²
- shear studs, ¾ in diameter; $F_u = 65$ kips/in²
- all columns are W10 × 33

- typical beams M1 and M2 are W16 × 26 (simply supported, fully composite with slab, shored construction)

- typical girders M3 are W21 × 44 (non-composite), with fully restrained moment connections

- service loads on typical beams M1 (first and second floors), $w_D = 750$ lbf/ft (all-inclusive); $w_L = 1000$ lbf/ft (non-reducible)

- service loads on typical beams M2 (first and second floors), $w_D = 375$ lbf/ft (all-inclusive); $w_L = 500$ lbf/ft (non-reducible)

- girder M3 supports only reactions from beam members M1

- elevations show shear diagram and moment diagram for end bay A–B on column line 1 due to two unit loads of 1000 lbf applied at third points

- use IBC load combination Eq. 16-2 (LRFD) or Eq. 16-9 (ASD)

- neglect lateral loads and lateral translation of frame

42. (a) Determine the required number of $\frac{3}{4}$ in diameter shear studs (assume one strong stud per rib) for full composite action of typical beam M1. Verify that the flexural strength is adequate, and check if the beam meets the live load deflection limit of $l/360$.

42. (b) Design the beam-to-girder connection for typical beam member M1. Use a single-plate connection. Draw the connection detail.

42. (c) Determine the design axial compression load and the design flexural moment for the first story column A-1 using the unit load diagrams in illustration II and illustration III. Include consideration of $P\text{-}\delta$ effects using AISC 360 App. 8. Ignore the self-weight of the members.

42. (d) Determine the effective length factor, K_x, for the first story column A-1 using the alignment charts in the AISC 360 App. 7 Commentary. Check the adequacy of the column using the critical interaction equation for compression and flexure. The column can be considered pinned at the base. The column flanges are laterally braced at the footing and at the first floor. Assume $M_{ry} = 0$ ft-kips and $K_y = 1.0$.

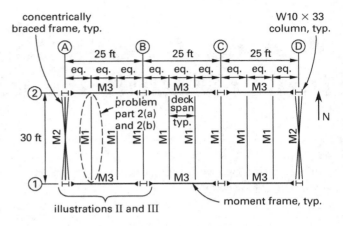

I. first floor and second floor plan

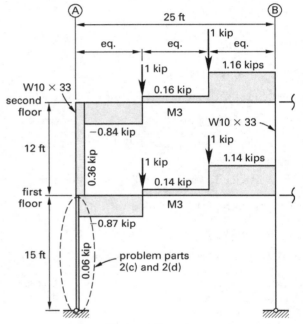

II. shear diagram for unit loads

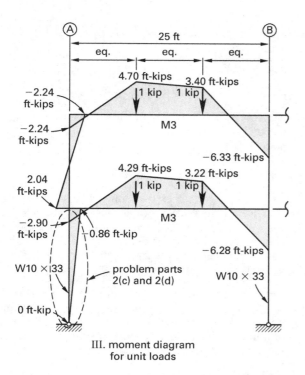

III. moment diagram
for unit loads

43. A 10 in, solid grouted, CMU cantilever retaining wall with a concrete footing is shown in illustration I. Illustration II shows the factored soil bearing pressure and factored vertical loads.

Design Codes

- ACI 318
- IBC
- TMS 402

Design Criteria

- masonry strength, $f'_m = 1500 \text{ lbf/in}^2$
- concrete strength, $f'_c = 4000 \text{ lbf/in}^2$
- reinforcing strength, $f_y = 60 \text{ kips/in}^2$
- masonry weight per surface area, $w_m = 105 \text{ lbf/ft}^2$
- concrete density, $\gamma_c = 150 \text{ lbf/ft}^3$
- soil density, $\gamma_s = 115 \text{ lbf/ft}^3$
- friction coefficient between concrete and soil, $\mu = 0.35$

- coefficient of active soil pressure, $K_a = 0.42$
- coefficient of at-rest soil pressure, $K_o = 0.59$
- coefficient of passive soil pressure, $K_p = 2.39$
- no hydrostatic pressure on wall
- do not consider wind or seismic forces

43. (a) Determine the factor of safety against sliding for the retaining wall. Include a sketch showing all horizontal forces acting on the wall. Do not consider the loads shown in illustration II.

43. (b) Calculate the axial force, shear force, and moment at the base of the CMU stem. Use ASD and IBC Eq. 16-9 $(D + H)$. Check the adequacy of the CMU. Do not consider the loads shown in illustration II.

43. (c) Using the factored soil bearing pressure and factored vertical loads shown in illustration II, check the adequacy of the concrete heel for shear and flexure. For flexure, assume no. 4 bars at 6 in o.c.

43. (d) The toe of the footing is also reinforced with no. 4 bars. Determine the required development length for the footing flexural reinforcement. Sketch a section through the reinforced footing that shows bar lengths and concrete clear cover. Do not detail the dowels for the CMU wall.

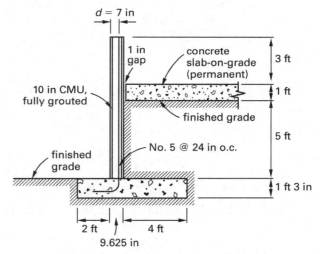

I. CMU retaining wall with a concrete footing

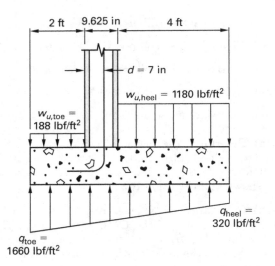

II. factored soil bearing pressure and factored vertical loads (enlarged)

Notes:
1. The shown loads and bearing pressures have been factored per IBC Eq. 16-2, $1.2D + 1.6H$.
2. $W_{u,toe}$ and $W_{u,heel}$ include all applicable vertical loads acting on the footing (e.g., soil, surcharge, and footing self-weight).

44. Illustration I shows a partial plan of a minor storage facility building in Fairbanks, Alaska. An elevation of a typical sawn lumber roof truss is shown in illustration II. The roof truss can be analyzed as an idealized, statically determinate truss: all members are pinned at their ends, and loads are only applied at the joints.

Design Codes

- ASCE/SEI7
- IBC
- NDS

Design Criteria

- low risk to human life in the event of failure
- roof dead load, $q_{dead} = 30 \text{ lbf/ft}^2$ (on inclined roof surface area)
- ground snow load, $p_g = 60 \text{ lbf/ft}^2$
- roof partially exposed, in terrain category C
- structure unheated with unobstructed, slippery roof surface
- sawn lumber species and grade, Douglas fir-larch no. 1 (see "Design Values")

- given adjustment factors, $C_M = C_t = C_i = 1.0$
- use ASD and consider dead load and balanced sloped roof snow load only per IBC Eq. 16-10 ($D + S$)
- ignore self-weight of truss members
- neglect offset between centerline of bearing on masonry wall and pinned connection at node 1 and node 5

44. (a) Determine the sloped roof snow load.

44. (b) Calculate the truss joint loads at nodes 1 through 5 and the truss vertical reactions at node 1 and node 5. Consider the balanced, sloped roof snow load calculated in part 44(a) and the roof dead load. Sketch an elevation of the truss, showing the joint loads and the truss reactions.

44. (c) Using the joint loads from part 44(b), calculate the axial force in member 1–6, and determine the smallest adequate size for this member. Assume the member has a nominal thickness of 3 in. Consider only the axial load away from the connections.

44. (d) For the bearing detail shown in illustration III, determine the required bearing length for member 1–2 at the top of the perimeter masonry wall for a 15 kip service level ($D + S$) truss reaction. Do not use the truss reaction calculated in part 44(b), and consider only the wood bearing stresses.

44. (e) The connection at node 6 is shown in illustration IV. Determine the capacity of the connection for a tensile force in member 2–6. Ignore the size calculated for member 1–6 in part 44(c), and do not consider local member stresses per NDS App. E. Assume $C_D = C_g = 1.0$.

Design Values: Douglas Fir-Larch No. 1 (Softwood)

- $F_b = 1000 \text{ lbf/in}^2$
- $F_t = 675 \text{ lbf/in}^2$
- $F_V = 180 \text{ lbf/in}^2$
- $F_{c\perp} = 625 \text{ lbf/in}^2$
- $F_c = 1500 \text{ lbf/in}^2$
- $E = 1{,}700{,}000 \text{ lbf/in}^2$
- $E_{min} = 620{,}000 \text{ lbf/in}^2$

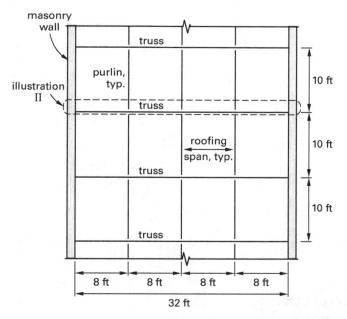

I. building plan

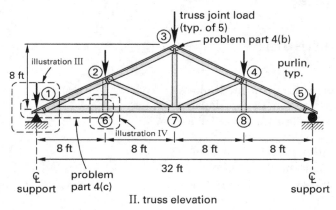

II. truss elevation

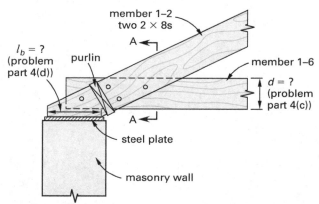

Note: Lateral bracing of truss at top of wall and tie-down straps not shown for clarity.

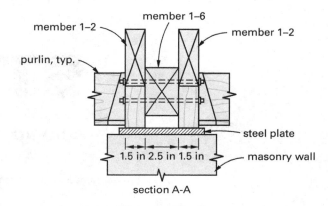

III. bearing detail

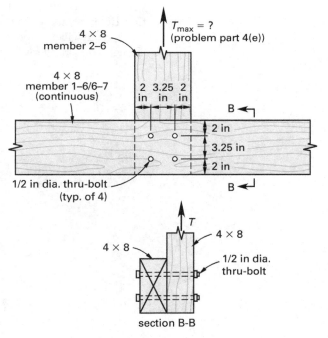

IV. node 6 connection

STOP!

DO NOT CONTINUE!

This concludes the Vertical Forces Component: Buildings Depth Module of the examination. If you finish early, check your work and make sure that you have followed all instructions. After checking your answers, you may turn in your Examination Booklet and Solution Booklet and leave the examination room. Once you leave, you will not be permitted to return to work or change your answers.

Lateral Forces Component: Breadth Module Instructions

In accordance with the rules established by your state, you may use textbooks, handbooks, bound reference materials, and any approved battery- or solar-powered, silent calculator to work this examination. However, no blank papers, writing tablets, unbound scratch paper, or loose notes are permitted. Sufficient room for scratch work is provided in the Examination Booklet.

You are not permitted to share or exchange materials with other examinees.

You will have four hours in which to work this module of the examination. Your score will be determined by the number of problems that you answer correctly. There is a total of 40 problems. There are no optional problems. Each problem is worth one point, and there is no penalty for incorrect answers. The maximum possible score for this section of the examination is 40 points.

Partial credit is not available. No credit will be given for methodology, assumptions, or work written in your examination booklet.

Record all of your answers on the Answer Sheet. No credit will be given for answers marked in the Examination Booklet. Mark your answers with a no. 2 pencil. Answers marked in pen may not be graded correctly. Marks must be dark and must completely fill the bubbles. Record only one answer per problem. If you mark more than one answer, you will not receive credit for the problem. If you change an answer, be sure the old bubble is erased completely; incomplete erasures may be misinterpreted as answers.

If you finish early, check your work and make sure that you have followed all instructions. After checking your answers, you may turn in your Examination Booklet and Answer Sheet and leave the examination room. Once you leave, you will not be permitted to return to work or change your answers.

When permission has been given by your proctor, break the seal on the Examination Booklet. Check that all pages are present and legible. If any part of your Examination Booklet is missing, your proctor will issue you a new Booklet.

Note that this practice exam is not an exact representation of the actual exam. For the exam, you must include your name, date of birth, examinee number, test site, and other information on a Scantron sheet. Each question will also be on a separate page with ample blank space for scratch work.

WAIT FOR PERMISSION TO BEGIN

Name: _____
      Last   First   Middle Initial

Examinee number: _____

Examination Booklet number: _____

Structural Engineering (SE) Examination

Lateral Forces Component Breadth Module

Lateral Forces Component: Breadth Module Answer Sheet

45. (A) (B) (C) (D)   55. (A) (B) (C) (D)   65. (A) (B) (C) (D)   75. (A) (B) (C) (D)
46. (A) (B) (C) (D)   56. (A) (B) (C) (D)   66. (A) (B) (C) (D)   76. (A) (B) (C) (D)
47. (A) (B) (C) (D)   57. (A) (B) (C) (D)   67. (A) (B) (C) (D)   77. (A) (B) (C) (D)
48. (A) (B) (C) (D)   58. (A) (B) (C) (D)   68. (A) (B) (C) (D)   78. (A) (B) (C) (D)
49. (A) (B) (C) (D)   59. (A) (B) (C) (D)   69. (A) (B) (C) (D)   79. (A) (B) (C) (D)
50. (A) (B) (C) (D)   60. (A) (B) (C) (D)   70. (A) (B) (C) (D)   80. (A) (B) (C) (D)
51. (A) (B) (C) (D)   61. (A) (B) (C) (D)   71. (A) (B) (C) (D)   81. (A) (B) (C) (D)
52. (A) (B) (C) (D)   62. (A) (B) (C) (D)   72. (A) (B) (C) (D)   82. (A) (B) (C) (D)
53. (A) (B) (C) (D)   63. (A) (B) (C) (D)   73. (A) (B) (C) (D)   83. (A) (B) (C) (D)
54. (A) (B) (C) (D)   64. (A) (B) (C) (D)   74. (A) (B) (C) (D)   84. (A) (B) (C) (D)

Lateral Forces Component: Breadth Module Exam

45. The enclosed simple diaphragm building with a parapet is subjected to wind loads.

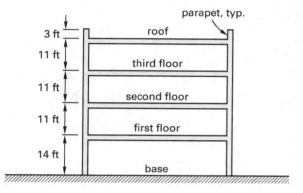

Design Code

- ASCE/SEI7

Design Criteria

- exposure category C
- internal pressure coefficient, $GC_{pi} = 0$ (solid parapet)
- risk category II
- topographical factor, $K_{zt} = 1.0$
- effective wind area, 10 ft^2
- basic wind speed (3 sec gust), v = 130 mi/hr
- use simplified procedure for components and cladding per ASCE/SEI7 Sec. 30.7.1.2

The total pressure on the corner zone of the windward parapet (load case A) is most nearly

- (A)   36 lbf/ft^2
- (B)   93 lbf/ft^2
- (C)   180 lbf/ft^2
- (D)   210 lbf/ft^2

46. A concrete column of a parking garage is subjected to the unfactored axial loads listed.

Design Codes

- ASCE/SEI7
- IBC

Service Level Loads (unfactored)

- axial dead load, $D = 200$ kips
- axial live load, $L = 75$ kips
- axial load due to horizontal seismic forces, $Q_E = 25$ kips

Design Criteria

- column, seismic design category E
- design spectral response acceleration parameter at 1 sec period, $S_{D1} = 0.75$
- design spectral response acceleration parameter at short period, $S_{DS} = 1.9$
- redundancy factor, $\rho = 1.3$

Using strength design load combinations, the maximum factored axial column load is most nearly

- (A)   350 kips
- (B)   360 kips
- (C)   380 kips
- (D)   420 kips

47. A free-standing bridge abutment is not restrained at the top by the bridge superstructure. The backfill material is cohesionless and unsaturated.

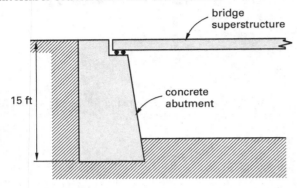

Design Code

- AASHTO

Design Criteria

- unit weight of soil, $\gamma_s = 0.130$ kips/ft^3
- seismic active pressure coefficient, $K_{AE} = 0.37$
- vertical seismic acceleration coefficient, $k_v = 0$
- horizontal seismic acceleration coefficient, $k_h = 0.10$
- backfill material internal friction angle, $\phi = 30°$

The total horizontal seismic active force, P_{AE}, exerted on the abutment is most nearly

- (A)   4.2 kips/ft
- (B)   4.4 kips/ft
- (C)   4.8 kips/ft
- (D)   5.4 kips/ft

48. A moment frame with columns pinned at the base is subjected to a lateral load, as shown.

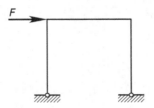

The deflected shape of the moment frame under the lateral load is

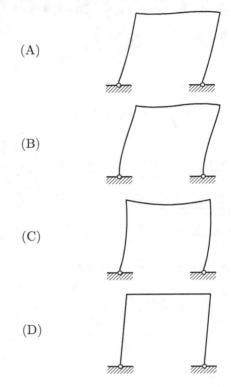

(A)

(B)

(C)

(D)

49. A police station will be built on a site that has a soil profile with three distinct soil layers. Each layer has a different shear wave velocity, as shown. The soil profile does not include any clay, peat, or liquefiable soils.

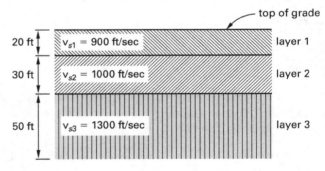

Design Code

- ASCE/SEI7

Design Criteria

- mapped spectral response acceleration at 0.2 sec, $S_S = 0.30$
- mapped spectral response acceleration at 1 sec, $S_1 = 0.080$
- the police station is an essential facility

The seismic design category for the police station is

(A) A

(B) B

(C) C

(D) D

50. A 25-story office building uses special steel moment frames for the seismic force-resisting system. Properties for the building are as given.

Design Code

- ASCE/SEI7

Design Criteria

- seismic weight, $W = 7500$ kips

- building height, $h_n = 300$ ft

- mapped spectral response acceleration at 0.2 sec, $S_S = 1.50$

- mapped spectral response acceleration at 1 sec, $S_1 = 0.50$

- mapped long-period transition period, $T_L = 8$ sec

- design spectral response acceleration at short period, $S_{DS} = 1.0$

- design spectral response acceleration at 1 sec period, $S_{D1} = 0.50$

Using the equivalent lateral force procedure, the seismic base shear is most nearly

(A) 180 kips

(B) 330 kips

(C) 450 kips

(D) 940 kips

51. A three-story structure is shown with each story's corresponding seismic weight.

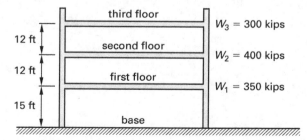

Design Code

- ASCE/SEI7

Design Criteria

- fundamental period of the structure, $T = 0.25$ sec

- seismic response coefficient, $C_s = 0.048$

Using the equivalent lateral force procedure, the seismic lateral force at the third floor is most nearly

(A) 13 kips

(B) 14 kips

(C) 21 kips

(D) 50 kips

52. A steel pipe is used to support a billboard. For seismic analysis, the billboard and supporting pipe are considered a nonbuilding structure. The self-weight of the pipe may be ignored.

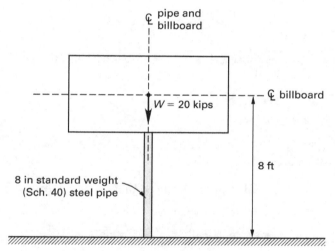

Design Codes

- AISC *Steel Construction Manual*

- ASCE/SEI7

Design Criteria

- design spectral response acceleration parameter at short period, $S_{DS} = 0.30$

- fundamental period of structure, $T = 0.046$ sec

- importance factor, $I = 1.0$

- weight of billboard, $W = 20$ kips

The maximum flexural stress in the pipe due to the strength level lateral seismic force is most nearly

(A)   2.5 kips/in^2

(B)   8.9 kips/in^2

(C)   11 kips/in^2

(D)   36 kips/in^2

53. A 40-story office building has no horizontal structural irregularities. Each line of seismic force-resisting members resists no more than 60% of the total seismic force.

Design Code

- ASCE/SEI7

Design Criteria

- mapped spectral response acceleration at 0.2 sec, $S_S = 1.70$

- mapped spectral response acceleration at 1 sec, $S_1 = 0.80$

- soil site class C

- risk category II

Which seismic force-resisting system could be used for the building?

(A)   special reinforced concrete moment frames

(B)   special reinforced concrete shear walls

(C)   steel eccentrically braced frames

(D)   all of the above

54. The elevation of a two-span, continuous concrete bridge is shown. The design response spectrum for the seismic analysis of the bridge is also provided.

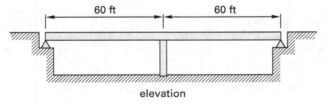

elevation

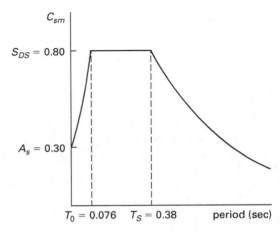

design response spectrum

Design Code

- AASHTO

Design Criteria

- average weight of bridge, $w = 7.5$ kips/ft

- lateral deflection at midspan of bridge, $v_{s,\max} = 0.0040$ ft due to uniform transverse lateral load, $p_o = 1.0$ kip/ft

Using the uniform load method, the equivalent uniform static transverse seismic load, p_e, is most nearly

(A)   5.0 kips/ft

(B)   6.0 kips/ft

(C)   7.5 kips/ft

(D)   8.1 kips/ft

55. An essential, two-span, continuous concrete bridge with a single-column central pier is shown. The column is fixed at the base and at the connection to the superstructure in both the transverse and longitudinal directions. The superstructure is restrained laterally only by the circular column. The column satisfies detailing requirements necessary for the use of the seismic response modification factor.

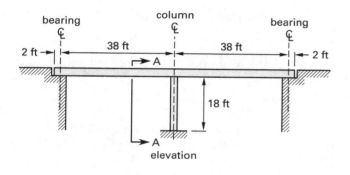

elevation

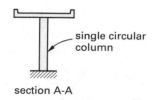

single circular column

section A-A

Design Code

- AASHTO

Design Criteria

- average weight of bridge, $w = 5.25$ kips/ft

- elastic seismic response coefficient, $C_{sm} = 0.50$

- period of vibration, $T_m = 0.22$ sec

The column moment due to earthquake loads is most nearly

(A)   950 ft-kips

(B)   990 ft-kips

(C)   1200 ft-kips

(D)   2000 ft-kips

56. The plan of a single-story, open-front concrete shear wall building is shown.

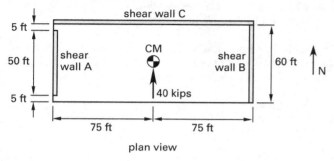

plan view

Design Code

- ASCE/SEI7

Design Criteria

- no horizontal or vertical structural irregularities

- roof diaphragm is rigid

- ignore out-of-plane stiffness and torsional stiffness of individual walls

- walls not connected at building corners

- relative rigidities of walls are proportional to their lengths

The shear force in wall A due to the shown seismic load is most nearly

(A)   17 kips

(B)   18 kips

(C)   20 kips

(D)   22 kips

57. A four-story, concrete building has special moment-resisting frames. The elastic story displacement for each story is shown. Assume that the lateral stiffness for each story is inversely proportional to the drift ratio of that story.

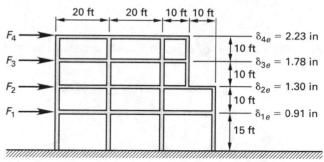

Design Code

- ASCE/SEI7

Which type of vertical structural irregularity exists for the building?

(A)   vertical geometric irregularity

(B)   stiffness-soft story irregularity

(C)   stiffness-extreme soft story irregularity

(D)   none of the above

58. The roof plan and section through a one-story building with a simply-supported, flexible roof diaphragm is shown.

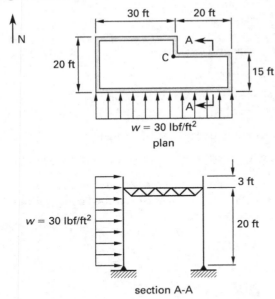

plan

section A-A

For the wind load, the unfactored chord force at point C is most nearly

(A) 6.1 kips

(B) 7.9 kips

(C) 8.2 kips

(D) 9.9 kips

59. The elevation of a carport with a pitched roof is shown. The carport is an open structure.

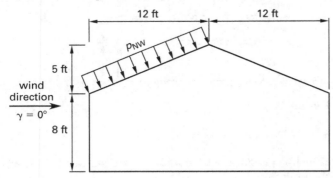

Design Code

- ASCE/SEI7

Design Criteria

- wind speed (3 sec gust), v = 170 mi/hr
- risk category II
- exposure category C

- topographical factor, $K_{zt} = 1.0$
- gust-effect factor, $G = 0.85$
- wind flow under roof is unobstructed (clear wind flow)
- do not consider load factors

For the design of the main wind force-resisting system, the net perpendicular wind pressure on the windward half of the roof surface, p_{NW}, is most nearly

(A) 50 lbf/ft^2

(B) 54 lbf/ft^2

(C) 59 lbf/ft^2

(D) 70 lbf/ft^2

60. A three-span bridge with simply supported girders is shown. The girders are restrained in the transverse direction at all supports.

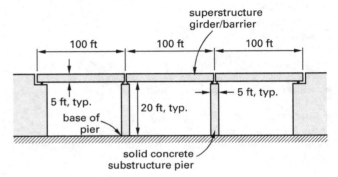

Design Code

- AASHTO

Design Criteria

- bridge height, $H < 30$ ft above ground
- design wind velocity, $v_{DZ} = 120$ mi/hr
- wind pressure acts perpendicular to the bridge span; skew angle, $S = 0°$

Considering wind load acting on the superstructure and substructure, the maximum unfactored transverse reaction at the base of one of the interior piers is most nearly

(A) 29 kips

(B) 36 kips

(C) 42 kips

(D) 49 kips

61. The elevation of a two-span, simply supported girder and slab bridge is shown. The girders are restrained in the transverse direction at all supports, and the pier is fixed at the base and free at the top (cantilever condition).

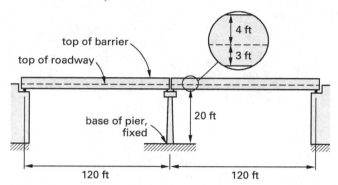

Design Code

- AASHTO

Design Criteria

- bridge height, $H < 30$ ft above ground

- design wind velocity, $v_{DZ} = v_B = 100$ mi/hr

- wind pressure acts perpendicular to the bridge span; skew angle, $S = 0°$

The maximum unfactored transverse moment at the base of the central pier due to wind pressure on vehicles is most nearly

- (A)   240 ft-kips
- (B)   310 ft-kips
- (C)   350 ft-kips
- (D)   400 ft-kips

62. A moment frame is subjected to the lateral loads shown. The columns are fixed at the base.

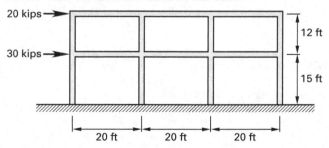

Using the portal method of simplified analysis, the moment at the base of an interior column is most nearly

- (A)   0 ft-kips
- (B)   94 ft-kips
- (C)   130 ft-kips
- (D)   250 ft-kips

63. A 15-story hospital building (risk category IV) with special steel concentrically braced frames is analyzed using the equivalent lateral force procedure. The elastic story deflections under strength-level earthquake forces are summarized in the computer output table provided.

floor level	height of story below	elastic deflection (relative to building base)
1	15 ft	0.45 in
2	12 ft	0.72 in
3	12 ft	0.99 in
4	15 ft	1.29 in
5	12 ft	1.69 in
6	12 ft	2.00 in
7	12 ft	2.29 in
8	12 ft	2.59 in
9	12 ft	2.90 in
10	12 ft	3.21 in
11	12 ft	3.52 in
12	12 ft	3.82 in
13	12 ft	4.14 in
14	10 ft	4.39 in
15	10 ft	4.68 in

Design Code

- ASCE/SEI7

At the tenth story (the story between the ninth and tenth floors), the ratio of the computed story drift to the allowable story drift is most nearly

- (A)   0.65
- (B)   0.72
- (C)   0.86
- (D)   1.10

64. A single-story, wood-framed building is shown. The exterior walls are simply supported at the ground and at the roof diaphragm. The roof framing members are not continuous at the collector element.

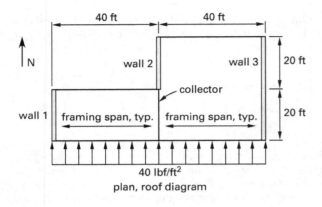

plan, roof diagram

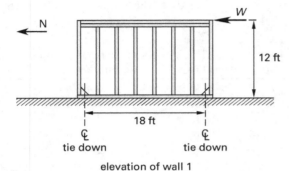

elevation of wall 1

Design Code

- IBC

Design Criteria

- flexible roof diaphragm with north-south lateral wind load, $q_{\text{wind}} = 40 \text{ lbf/ft}^2$

- roof dead load, $q_{\text{dead}} = 15 \text{ lbf/ft}^2$

- roof live load, $q_{\text{live}} = 30 \text{ lbf/ft}^2$

- roof snow load, $q_{\text{snow}} = 35 \text{ lbf/ft}^2$

- no lateral loads due to earth pressure

- ignore wall self-weight

Using the controlling LRFD load combination, the maximum tensile force in the tie-down of shear wall 1 is most nearly

(A) 200 lbf

(B) 500 lbf

(C) 1400 lbf

(D) 3200 lbf

65. A section through a building in seismic design category D is shown. The building has tilt-up concrete walls and a flexible roof diaphragm. Nailed steel straps at 4 ft o.c. are used to anchor the walls for out-of-plane earthquake loads.

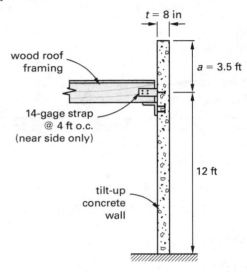

Design Codes

- ASCE/SEI7

- IBC

Design Criteria

- concrete density, $\gamma_c = 150 \text{ lbf/ft}^3$

- design spectral response acceleration parameter at short period, $S_{DS} = 0.60$

- span of flexible diaphragm, $L_f = 100 \text{ ft}$

- occupancy importance factor, $I_e = 1.0$

The strength design tensile force in the steel strap is most nearly

(A) 1100 lbf

(B) 1600 lbf

(C) 1800 lbf

(D) 2600 lbf

66. The elevation of a two-span, steel girder highway bridge in seismic zone 3 is shown. The supports are not skewed relative to the bridge span.

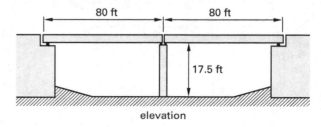

elevation

Design Code

- AASHTO

The minimum support length at the bridge abutment is most nearly

- (A)   11 in
- (B)   12 in
- (C)   13 in
- (D)   17 in

67. According to the AISC *Seismic Construction Manual*, which statement about steel eccentrically braced frames (EBFs) is INCORRECT?

- (A)   EBFs provide ductility primarily through out-of-plane buckling of the gusset plates and through yielding/buckling of the braces.

- (B)   EBFs can often fit in locations within the architectural floor plan where concentrically braced frames cannot.

- (C)   EBFs have higher response modification factors and lower seismic base shears than concentrically braced frames.

- (D)   EBFs limit nonstructural damage better than buildings with steel moment frames.

68. A beam-to-column moment connection is made with bolted flange plates, as shown.

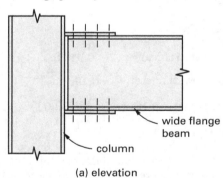

(a) elevation

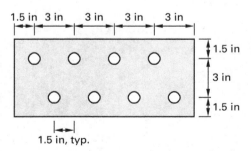

(b) view, top of flange plate

Design Code

- AISC 360

Design Criteria

- plate steel, A572 grade 42; $F_u = 60$ kips/in^2; $F_y = 42$ kips/in^2

- bolts, $\frac{3}{4}$ in diameter with standard holes

- plate thickness, $\frac{1}{2}$ in

Ignoring the block shear in the plate, what is most nearly the allowable tensile strength of the $\frac{1}{2}$ in plate? (LRFD answers for design tensile strength are given in parentheses.)

- (A)   63 kips (95 kips)
- (B)   67 kips (100 kips)
- (C)   70 kips (106 kips)
- (D)   75 kips (110 kips)

69. A steel special moment frame with reduced beam section moment connections is shown.

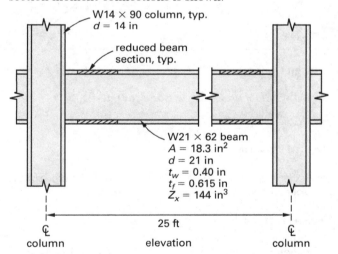

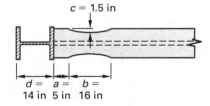

plan of reduced beam

Design Codes

- AISC *Seismic Construction Manual*
- AISC 341
- AISC 358

Design Criteria

- beam steel, ASTM A570 grade 55; $F_u = 65$ kips/in^2; $F_y = 55$ kips/in^2; $R_y = 1.1$

The probable maximum moment, M_{pr}, at the center of the reduced beam section is most nearly

- (A) 540 ft-kips
- (B) 590 ft-kips
- (C) 640 ft-kips
- (D) 790 ft-kips

70. An eccentrically braced frame with a 42 in shear link is shown. All connections are fixed.

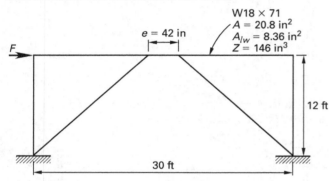

W18 × 71
$A = 20.8$ in^2
$A_{lw} = 8.36$ in^2
$Z = 146$ in^3

$e = 42$ in

F

12 ft

30 ft

Design Codes

- AISC *Seismic Construction Manual*
- AISC 341

Design Criteria

- axial load in the W18 × 71 link under the controlling seismic load combination, $P_u = 150$ kips
- steel strength, $F_y = 50$ kips/in^2

The nominal shear strength of the shear link is most nearly

- (A) 190 kips
- (B) 250 kips
- (C) 350 kips
- (D) 620 kips

71. The plan for an existing building with clay masonry perimeter walls and wood joists is shown. To retrofit the diaphragm, 18-gage metal straps are added below the existing floor to resist the shown service-level lateral wind load. The straps are spliced using no. 10 screws. The shear strength of the connection is not limited by end distance.

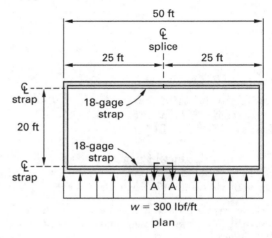

50 ft

CL splice

25 ft   25 ft

CL strap

18-gage strap

20 ft

18-gage strap

CL strap

A A

w = 300 lbf/ft

plan

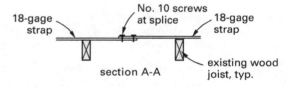

18-gage strap

No. 10 screws at splice

18-gage strap

section A-A

existing wood joist, typ.

Design Code

- *AISI Specification*

Design Criteria

- 18-gage strap thickness, $t = 0.048$ in
- no. 10 screw diameter, $d = 0.190$ in
- tensile strength of straps, $F_u = 65{,}000$ lbf/in^2
- use ASD, $\Omega = 3.0$

At the centerline of the building, the minimum number of no. 10 screws required at a strap splice is most nearly

- (A) 3 screws
- (B) 4 screws
- (C) 10 screws
- (D) 12 screws

72. A reinforced coupling beam that is part of an 18 in thick, special concrete shear wall is shown.

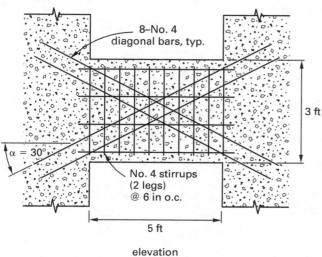

elevation

Design Code

- ACI 318

Design Criteria

- concrete strength, $f_c' = 4000$ lbf/in^2

- steel strength, $f_y = 60$ kips/in^2

- transverse reinforcement for diagonal bars not shown for clarity

The nominal shear capacity of the coupling beam is most nearly

- (A) 79 kips
- (B) 96 kips
- (C) 190 kips
- (D) 410 kips

73. A beam is part of an intermediate concrete moment frame, as shown.

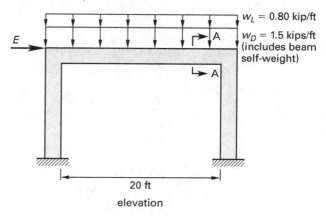

elevation

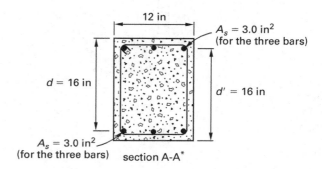

section A-A*

*Reinforcement is as shown for the full length of the beam.

Design Code

- ACI 318

Design Criteria

- concrete strength, $f_c' = 4000$ lbf/in^2

- shear at beam end due to unfactored earthquake loads, $V_{EQ} = 8.0$ kips

- neglect axial load in beam

- steel reinforcement strength, $f_y = 60$ kips/in^2

Using ACI 318 Eq. 5.3.1e (load combination $1.2D + 1.0E + 1.0L$), the design shear force at the beam end is most nearly

- (A) 31 kips
- (B) 34 kips
- (C) 42 kips
- (D) 47 kips

74. A column is part of an ordinary, non-sway, concrete moment frame. The factored axial load and factored end moments in the column (from a first-order analysis) are shown.

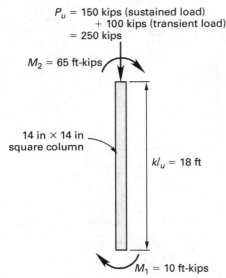

Design Code

- ACI 318

Design Criteria

- column flexural stiffness (already reduced for cracking, creep, and nonlinearity), $EI = 2.885 \times 10^6$ in²-kips
- concrete strength, $f_c' = 4000$ lbf/in²

The maximum amplified factored moment for the column's design is most nearly

(A)   65 ft-kips

(B)   77 ft-kips

(C)   82 ft-kips

(D)   95 ft-kips

75. A cast-in headed stud subjected to a shear load is shown.

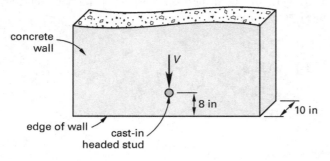

Design Code

- ACI 318

Design Criteria

- anchor design governed by concrete breakout (nonductile)
- basic concrete breakout strength in shear of a single anchor, $V_b = 11.5$ kips
- concrete cracked, no supplementary reinforcement
- given modification factors, $\Psi_{ed,V} = \Psi_{c,V} = 1.0$

The nominal concrete breakout strength, V_{cb}, is most nearly

(A)   10.5 kips

(B)   11.1 kips

(C)   11.5 kips

(D)   12.6 kips

76. A transverse section of a bridge pier is shown.

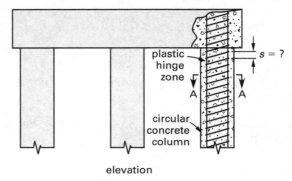

elevation

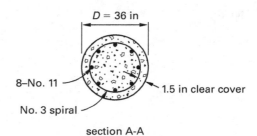

section A-A

Design Code

- AASHTO

Design Criteria

- concrete strength, $f_c' = 4$ kips/in²
- steel reinforcement strength, $f_y = f_{yh} = 60$ kips/in²

- seismic zone 4

- design of spiral reinforcement not governed by shear and torsion

The maximum permitted spacing (pitch), s, of the no. 3 spiral reinforcement in a plastic hinge region is most nearly

- (A) 2.0 in
- (B) 4.0 in
- (C) 6.0 in
- (D) 8.5 in

77. An unblocked wood structural panel diaphragm that is subjected to a lateral wind load is shown.

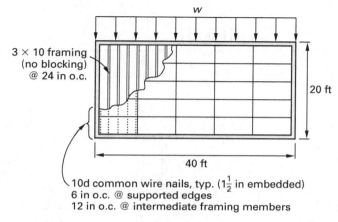

Design Codes

- NDS
- SDPWS

Design Criteria

- framing, spruce-pine-fir
- panel, ½ in structural I

Using ASD, the allowable diaphragm shear is most nearly

- (A) 240 lbf/ft
- (B) 310 lbf/ft
- (C) 620 lbf/ft
- (D) 670 lbf/ft

78. A flexible roof diaphragm of a single-story, wood-framed building is shown. The diaphragm is subjected to a lateral diaphragm-level wind load of 200 lbf/ft.

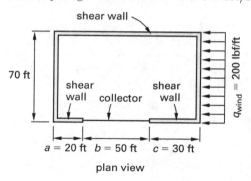

The maximum axial force in the collector (drag strut) element is most nearly

- (A) 1400 lbf
- (B) 2100 lbf
- (C) 3500 lbf
- (D) 7000 lbf

79. A structure built using a concrete masonry unit (CMU) wall with a precast concrete plank roof is shown. The wall is subjected to an out-of-plane wind load.

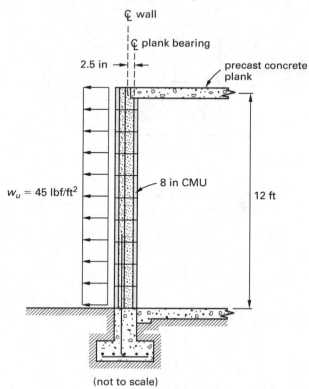

Design Code

- TMS 402

Design Criteria

- total factored axial load at mid-height of wall, $P_u = 2300$ lbf/ft (includes contributions from wall and roof areas)

- factored load from tributary roof area, $P_{uf} = 1600$ lbf/ft

- horizontal mid-height deflection of wall, $\delta = 0.08$ in

- assume the wall is simply supported for out-of-plane loads

The factored design moment, M_u, at the mid-height of the wall is most nearly

(A) 250 ft-lbf/ft

(B) 810 ft-lbf/ft

(C) 970 ft-lbf/ft

(D) 990 ft-lbf/ft

80. The elevation of a fully grouted, special reinforced masonry shear wall is shown.

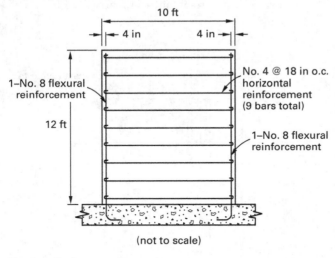

(not to scale)

Design Code

- TMS 402

Design Criteria

- wall thickness, 10 in (nominal)

- masonry laid in running bond

- seismic design category D

- use ASD

To meet the requirements of a special reinforced masonry shear wall, the required number of additional, equally spaced, vertical no. 4 reinforcing bars between the no. 8 flexural bars is most nearly

(A) 2 bars

(B) 3 bars

(C) 5 bars

(D) 7 bars

81. A headed anchor bolt is embedded into the top of a fully grouted, 6 in CMU wall, as shown. The anchor is subjected to an axial tensile force due to suction wind load.

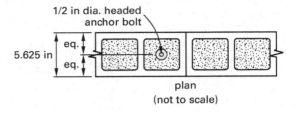

plan
(not to scale)

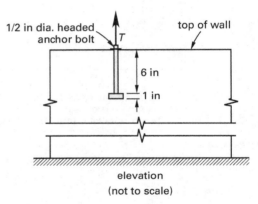

elevation
(not to scale)

Design Code

- TMS 402

Design Criteria

- anchor bolts, $\frac{1}{2}$ in diameter ASTM F1554 grade 55, $f_y = 55$ kips/in^2

- masonry strength, $f'_m = 1500$ lbf/in^2

- anchor's projected tension area on the masonry surface of a right circular cone, $A_{pt} = 64.9$ in^2

- use ASD

The allowable axial tensile load of the headed anchor bolt is most nearly

(A) 0.48 kip

(B) 3.1 kips

(C) 5.5 kips

(D) 6.5 kips

82. A concrete bridge pier with concrete piles founded on hard rock is shown in elevation and plan view.

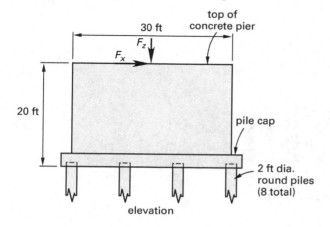

elevation

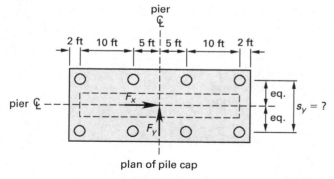

plan of pile cap

Design Code

- AASHTO

Design Criteria

- all factored loads act concurrently

- transverse force at top of pier, $F_x = 50$ kips

- longitudinal force at top of pier, $F_y = 30$ kips

- vertical load (includes self-weight of pier and pile cap plus superstructure load), $F_z = 500$ kips

- axial pile capacity for factored loads, $P = 100$ kips

- pile cap is rigid

Considering the axial pile capacity, the minimum required longitudinal center-to-center spacing of the piles is most nearly

(A) 3 ft 4 in

(B) 5 ft 0 in

(C) 6 ft 8 in

(D) 7 ft 9 in

83. The rectangular spread footing shown is subjected to an axial load and applied moments in two orthogonal directions. The axial load includes an allowance for the column and footing self-weights.

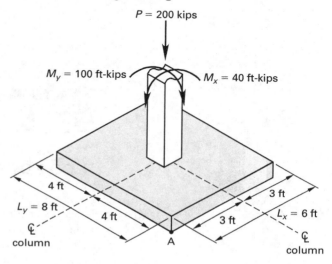

The maximum bearing pressure at point A is most nearly

(A) 3.3 tons/ft^2

(B) 3.5 tons/ft^2

(C) 3.8 tons/ft^2

(D) 4.2 tons/ft^2

84. According to the IBC, which of the following buildings does NOT require structural observations?

(A) an emergency shelter (risk category IV) in seismic design category D

(B) a 50 ft tall office building (risk category II) in seismic design category E

(C) a 60 ft tall office building (risk category II) in seismic design category D

(D) a 100 ft tall office building (risk category II) located where the ASD wind velocity is 120 mph

STOP!

DO NOT CONTINUE!

This concludes the Lateral Forces Component: Breadth Module of the examination. If you finish early, check your work and make sure that you have followed all instructions. After checking your answers, you may turn in your Examination Booklet and Answer Sheet and leave the examination room. Once you leave, you will not be permitted to return to work or change your answers.

Lateral Forces Component: Buildings Depth Module Instructions

In accordance with the rules established by your state, you may use textbooks, handbooks, bound reference materials, and any approved battery- or solar-powered, silent calculator to work this examination. However, no blank papers, writing tablets, unbound scratch paper, or loose notes are permitted. Sufficient room for showing your work is provided in the Solution Booklet.

You are not permitted to share or exchange materials with other examinees.

You will have four hours in which to work this module of the examination. Your solutions to the essay problems will be evaluated by two subject matter experts for overall compliance with established scoring criteria and general quality. A third subject matter expert will be used when necessary. There is a total of four problems. All four problems are equally weighted. They must be worked correctly in order to receive full credit on the exam. There are no optional problems.

Along with your Examination Booklet, a Solution Booklet of graph paper is provided for each problem. Record your solutions in the appropriate Booklet for each problem. No credit will be given for solutions written in the incorrect Booklet.

If you finish early, check your work and make sure that you have followed all instructions. After checking your answers, you may turn in your Examination and Solution Booklets and leave the examination room. Once you leave, you will not be permitted to return to work or change your answers.

When permission has been given by your proctor, break the seals on the Examination and Solution Booklets. Check that all pages are present and legible. If any part of your Examination or Solution Booklet is missing, your proctor will issue you a new Booklet.

Note that this practice exam is not an exact representation of the actual exam. For the exam, you must include your name, date of birth, examinee number, test site, and other information on a Scantron sheet. You will be provided four separate booklets of blank graph paper (one for each problem) in which to show your work. You must write your name and examinee number on each booklet.

Structural Engineering (SE) Examination

Lateral Forces Component Buildings Depth Module

Lateral Forces Component: Buildings Depth Module Exam

85. A single-story building with special reinforced masonry shear walls and a flexible roof diaphragm is shown.

Design Codes

- ASCE/SEI7
- IBC
- TMS 402

Design Criteria

- roof dead load, $q_{dead} = 25$ lbf/ft^2
- seismic weight tributary to roof level, $W = 600$ kips
- seismic importance factor, $I_e = 1.0$
- soil site class C
- mapped spectral response acceleration at short period, $S_S = 1.40$
- mapped spectral response acceleration at 1 sec, $S_1 = 0.53$
- mapped long-period transition period, $T_L = 8$ sec
- redundancy factor, $\rho = 1.0$
- ignore wind and snow loads
- ignore P-delta effects
- use ASD

Masonry Properties

- 10 in (nominal) concrete masonry units laid in running bond with $1\frac{1}{4}$ in face shells, cells grouted at 32 in o.c.
- net area, $A_n = 52.4$ in^2/ft
- net moment of inertia, $I_n = 624.6$ in^4/ft
- masonry strength, $f'_m = 1500$ lbf/in^2
- reinforcing strength, $f_y = 60$ kips/in^2
- masonry weight per surface area, $w_m = 74$ lbf/ft^2

85. (a) Using the equivalent lateral force procedure, determine the in-plane seismic shear load in wall C, considering the east-west seismic load at the roof level. Use the ASD load factor of $0.7E$.

85. (b) Using the seismic shear load calculated in part 85(a), determine the required spacing of no. 5 horizontal reinforcing bars in shear wall C. Assume that the wall has an effective depth of 19 ft. Neglect the axial load in the wall. Do not check the seismic reinforcing requirements, and do not include a one-third increase when calculating the allowable shear stress.

85. (c) Determine the design out-of-plane moment and axial load at the mid-height of shear wall A using ASD load combination $D + 0.7E$. Include the vertical seismic load effect.

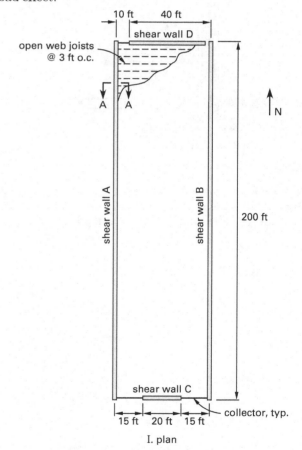

I. plan

85. (d) Using the design moment and axial load calculated in part 85(c), check the adequacy of no. 5 vertical bars at 32 in o.c. for combined compression and flexure at the mid-height of shear wall A.

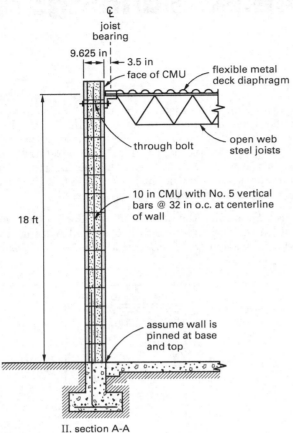

II. section A-A

86. The plan and elevation of a single-story, wood-frame building are shown. The building is enclosed, the roof is flat, and the roof diaphragm is flexible.

Design Codes

- ASCE/SEI7
- IBC
- NDS

Design Criteria

- basic wind speed, v = 140 mi/hr
- exposure category C
- gust-effect factor, $G = 0.85$
- given adjustment factors, $C_M = C_t = 1.0$
- sawn lumber species and grade is eastern softwoods no. 2
- use ASD

Shear Wall Construction Properties

- sheathing, wood structural panels (one side) nailed directly to studs
- wall studs, 2 in × 6 in at 24 in o.c.

86. (a) Determine the worst-case strength design wind pressure on a corner zone of the building's south wall. Use the envelope procedure (Part 1 of ASCE/SEI7 Chap. 28) for wind loads on the main wind force-resisting system of a low-rise building. Do not use the wind loads shown in plans III and IV.

86. (b) For the roof-level lateral wind load shown in plans III and IV, calculate the maximum axial tensile force in the double 2 in × 6 in top plate of the east wall. Consider both the east-west and north-south wind loads, but do not consider them to act simultaneously. Do not use the wind load calculated in part 86(a).

86. (c) For the axial load calculated in part 86(b), determine the number of 8d box nails required at a splice of the double top plate. Refer to illustration V.

86. (d) For the roof-level lateral wind load shown in plans III and IV, design the sheathing thickness and nailing pattern for shear wall A. Do not consider out-of-plane wind loads.

86. (e) Sketch the connection detail between the roof framing and shear wall A. Identify all required components, but do not design.

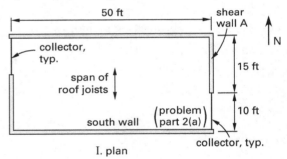

I. plan

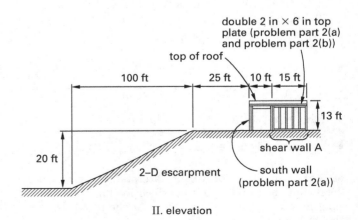

II. elevation

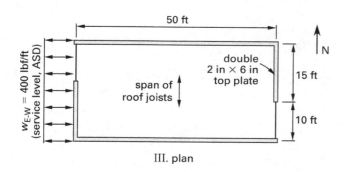

III. plan

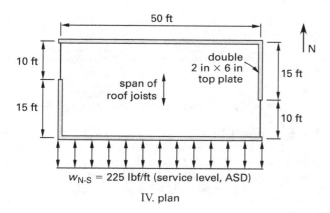

w_{N-S} = 225 lbf/ft (service level, ASD)

IV. plan

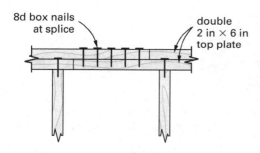

V. shear wall A splice detail

87. A typical floor plan of a multi-story, concrete building with rigid diaphragms is shown in illustration I. The lateral system consists of two four-bay special reinforced concrete moment frames in the north-south direction and two special reinforced concrete shear walls in the east-west direction.

Design Codes

- ACI 318
- IBC

Design Criteria

- concrete strength, $f'_c = 5000$ lbf/in² (normal weight)

- reinforcing strength, $f_y = 60,000$ lbf/in²

- rigidity of one of the shear walls is five times the rigidity of one of the four-bay moment frames

- total building (and shear wall) height, $h_w = 60$ ft

- building has no horizontal or vertical structural irregularities

- neglect contribution from compression steel for calculation of flexural strengths of beams

Distributed Shear Wall Reinforcement (away from boundary elements)

- two layers of no. 5 vertical bars at 12 in o.c.

- two layers of no. 6 horizontal bars at 8 in o.c.

87. (a) Determine the shear force in wall C due to a 500 kip strength-level seismic story force in the east-west direction.

87. (b) Determine the adequacy of the provided horizontal shear wall reinforcement for the shear force calculated in part 87(a). Check if the wall vertical and horizontal reinforcement ratios satisfy the requirements for a special reinforced concrete shear wall.

87. (c) For the moment frame joint shown in illustration II, determine whether the column meets the minimum flexural strength requirements of ACI 318 Sec. 18.7.3.2 Use illustration III and consider a factored axial load of 300 kips above the floor level and 350 kips below the floor level. Neglect axial load in the beam.

87. (d) For the end moment frame joint at gridline A-1, determine if a no. 9 longitudinal bar of the moment frame beam can be fully developed in tension through the column core with a standard 90° hook.

87. (e) Sketch an elevation of the end moment frame joint at gridline A-1. Show all required bars, but do not design.

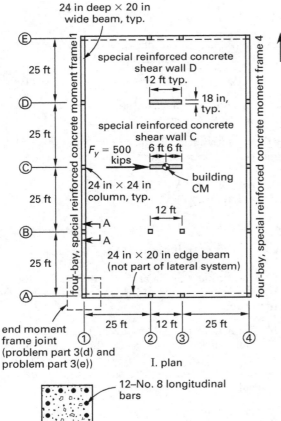

I. plan

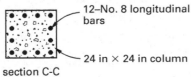

section C-C

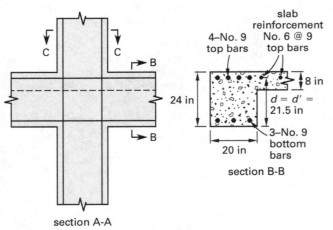

II. sections A-A, B-B, and C-C

Note: Transverse reinforcement not shown for clarity.

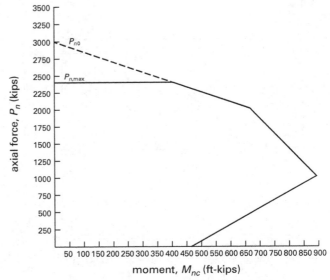

III. column interaction diagram
24 in × 24 in column with 12–No. 8 longitudinal bars

88. Illustration I shows a typical floor plan of a four-story, steel office building with special concentrically braced frames. An elevation of the inverted V (chevron) braced frame on gridline 1, with horizontal seismic forces for each floor level, is shown in illustration II.

Design Codes

- ACI 318
- AISC 341
- AISC *Steel Construction Manual*
- AISC 360
- AISC *Seismic Construction Manual*
- ASCE/SEI7

Steel

- columns and beams, ASTM A992 grade 50; $F_y = 50$ kips/in^2; $F_u = 65$ kips/in^2
- HSS braces, ASTM A500 grade B; $F_y = 46$ kips/in^2; $F_u = 58$ kips/in^2
- welding electrodes, $F_{EXX} = 70$ kips/in^2

Concrete Footing

- concrete strength, $f'_c = 4000$ lbf/in^2 (normal weight)
- reinforcing strength, $f_y = 60,000$ lbf/in^2

Loads

- dead load (floors 1–3 and roof), $q_{\text{dead}} = 150$ lbf/ft^2 (includes allowance for facade, column, and beam self-weight)

- live load (floors 1–3), $q_{\text{live}} = 50$ lbf/ft^2

- snow load (roof only), $q_{\text{snow}} = 30$ lbf/ft^2 (flat roof)

- horizontal seismic load effect on braced frame 1, see illustration II

- design spectral response acceleration at short periods, $S_{DS} = 0.40$

- redundancy factor, $\rho = 1.0$

- seismic design category C

- use the LRFD method and consider only ASCE/SEI7 Sec. 12.4.2.3, load combination 5, $(1.2 + 0.2S_{DS})D + \rho Q_E + 0.5L + 0.2S$

Analysis Assumptions

- all members are pinned at their ends

- use workpoint-to-workpoint dimensions

88. (a) Determine the maximum factored vertical footing reaction at gridline B-1. Consider live load reduction, and ignore the footing self-weight.

88. (b) For the reaction determined in part 4(a), check the adequacy of the spread footing (shown in illustration III) for punching shear.

88. (c) Using illustration II and the additional design criteria listed, determine the expected strengths in tension and compression of the HSS6 $\times$ 6 $\times$ $^3/_8$ in braces. Then determine the maximum moment demand on the W24 $\times$ 250 (beam 1).

Design Criteria for Part 4(c)

- distributed dead load on beam 1, $w_{\text{dead}} = 750$ lbf/ft (includes beam self-weight)

- distributed live load on beam 1, $w_{\text{live}} = 250$ lbf/ft (non-reducible)

88. (d) Describe how the braces could be reconfigured in order to reduce the moment demand on beam 1. Sketch an elevation of the braced frame on gridline 1 showing the reconfiguration.

88. (e) Determine the required size and length of the brace-to-gusset welds at node 1.

88. (f) The contractor submits a Request for Information (RFI) asking if it is permissible to install powder-actuated fasteners into the gusset plates of the brace-to-beam

connection in order to hang ductwork. Reference relevant code sections to respond to the contractor.

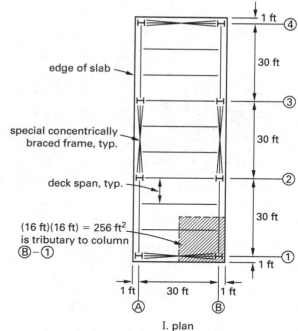

I. plan

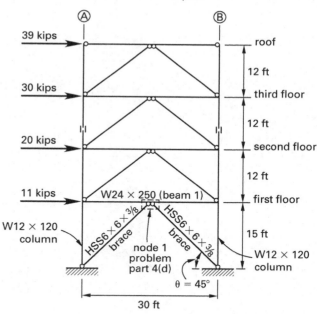

II. elevation, braced frame on gridline ①

III. elevation, column (B)–(1) footing detail

STOP!

DO NOT CONTINUE!

This concludes the Lateral Forces Component: Buildings Depth Module of the examination. If you finish early, check your work and make sure that you have followed all instructions. After checking your answers, you may turn in your Examination Booklet and Solution Booklet and leave the examination room. Once you leave, you will not be permitted to return to work or change your answers.

Answer Keys

Vertical Forces Component: Breadth Module Answer Key

1. C	11. D	21. C	31. B
2. D	12. C	22. C	32. B
3. C	13. A	23. A	33. D
4. B	14. C	24. A	34. B
5. D	15. A	25. D	35. A
6. D	16. B	26. C	36. C
7. D	17. D	27. B	37. B
8. A	18. C	28. B	38. D
9. D	19. C	29. C	39. D
10. C	20. A	30. C	40. C

Lateral Forces Component: Breadth Module Answer Key

45. C	55. B	65. D	75. A
46. D	56. D	66. D	76. A
47. D	57. B	67. A	77. B
48. A	58. B	68. B	78. B
49. C	59. A	69. B	79. D
50. B	60. C	70. B	80. A
51. C	61. C	71. D	81. B
52. C	62. C	72. B	82. C
53. A	63. B	73. C	83. A
54. B	64. B	74. B	84. C

Solutions
Vertical Forces Component: Breadth Module Exam

1. From the problem illustration and ASCE/SEI7 Fig. 7-8, l_u is 50 ft for leeward drift and 200 ft for windward drift. The clear height, h_c, is 10 ft.

From ASCE/SEI7 Sec. 7.7.1, the drift height, h_d, is calculated using ASCE/SEI7 Fig. 7-9. The controlling drift height, h_d, is the larger of the value from ASCE/SEI7 Fig. 7-9 considering l_u for leeward drift or $\frac{3}{4}$ of the value from Fig. 7-9 considering l_u for windward drift. (These equations are not dimensionally consistent.)

$$h_d = \max \begin{cases} 0.43\sqrt[3]{l_u}\sqrt[4]{p_g + 10} - 1.5 \\ \quad = (0.43)\sqrt[3]{50\text{ ft}}\sqrt[4]{60\ \dfrac{\text{lbf}}{\text{ft}^2} + 10\text{ ft}} - 1.5 \\ \quad = 3.08\text{ ft} < h_c \\ \left(\tfrac{3}{4}\right)\left(0.43\sqrt[3]{l_u}\sqrt[4]{p_g + 10} - 1.5\right) \\ \quad = \left(\dfrac{3}{4}\right)\left(\begin{array}{l}(0.43)\sqrt[3]{200\text{ ft}} \\ \times \sqrt[4]{60\ \dfrac{\text{lbf}}{\text{ft}^2} + 10\text{ ft}} - 1.5\end{array}\right) \\ \quad = 4.33\text{ ft} < h_c \quad \text{[controls]} \end{cases}$$

The snow density, γ, is found using ASCE/SEI7 Eq. 7.7-1 and cannot be greater than 30 lbf/ft³. (This equation is not dimensionally consistent.)

$$\gamma = 0.13p_g + 14 = (0.13)\left(60\ \frac{\text{lbf}}{\text{ft}^2}\right) + 14$$
$$= 21.8\text{ lbf/ft}^3 < 30\text{ lbf/ft}^3 \quad \text{[OK]}$$

From ASCE/SEI7 Sec. 7.7.1, the maximum intensity for the drift surcharge load, p_d, is

$$p_d = h_d\gamma = (4.33\text{ ft})\left(21.8\ \frac{\text{lbf}}{\text{ft}^3}\right) = 94\text{ lbf/ft}^2$$

The answer is (C).

2. From AASHTO Sec. 3.6.1.2.1, the HL-93 live load consists of a design truck or tandem and a design lane load.

The design lane load, w, is given in AASHTO Sec. 3.6.1.2.4 as 0.64 kips/ft. The maximum moment for the design lane load, M_{lane}, is

$$M_{\text{lane}} = \frac{wl^2}{8}$$
$$= \frac{\left(0.64\ \dfrac{\text{kips}}{\text{ft}}\right)(30\text{ ft})^2}{8}$$
$$= 72\text{ ft-kips}$$

For single spans of less than 40 ft, the design tandem loading of AASHTO Sec. 3.6.1.2.3 will produce larger moment demands than the design truck loading. Therefore, use two 25 kip axle loads, P, with a longitudinal spacing, a, of 4 ft.

The moment due to the design tandem can be found from AISC *Steel Construction Manual* Table 3-23, case 44.

Determine whether a is greater than or less than $0.586l$.

$$0.568l = (0.568)(30\text{ ft}) = 17\text{ ft} > 4\text{ ft}$$

So, for $a < 0.568l$, the maximum moment for the design tandem, M_{tandem}, is

$$M_{\text{tandem}} = \left(\frac{P}{2l}\right)\left(l - \frac{a}{2}\right)^2$$
$$= \left(\frac{25\text{ kips}}{(2)(30\text{ ft})}\right)\left(30\text{ ft} - \frac{4\text{ ft}}{2}\right)^2$$
$$= 327\text{ ft-kips}$$

From AASHTO Table 3.4.1-1, live loads must be factored by 1.75 for the strength I limit state.

From AASHTO Sec. 3.6.2.1, the design tandem moment (but not the lane load) must be increased by the dynamic load allowance, IM, specified in AASHTO Table 3.6.2.1-1. For the design tandem moment, which falls under "all other limit states," IM is 33%.

Although the maximum moments due to the lane load and tandem load do not occur in the same location, it is sufficiently accurate to add the lane and tandem moments to get the maximum factored design live load moment, M_{max}.

$$M_{max} = 1.75\big(M_{lane} + (1 + IM)M_{tandem}\big)$$
$$= (1.75)\big(72 \text{ ft-kips} + (1 + 0.33)(327 \text{ ft-kips})\big)$$
$$= 887 \text{ ft-kips} \quad (890 \text{ ft-kips})$$

The answer is (D).

3. From AASHTO Comm. Eq. C5.4.2.4-1, the modulus of elasticity, E_c, for normal weight concrete is

$$E_c = 1820\sqrt{f_c'} = (1820)\sqrt{3 \ \frac{\text{kips}}{\text{in}^2}}$$
$$= 3152 \text{ kips/in}^2$$

From AASHTO Sec. 5.4.2.2, the thermal coefficient of expansion, α, for normal weight concrete is

$$\alpha = 6.0 \times 10^{-6}/°\text{F}$$

From AASHTO Table 3.4.1-1, the load factor for force effects due to uniform temperature is listed as 1.00 or 1.20. AASHTO Sec. 3.4.1 states that the larger of the two values must be used for deformation calculations, and the smaller value must be used for other effects. This problem requires calculating a force, not a deformation, so use $\gamma_{TU} = 1.00$.

The force resulting from restraint of thermal expansion is equal to the modulus of elasticity multiplied by the coefficient of expansion, multiplied by the temperature range, multiplied by the area of the member. Equations for thermal expansion can be found in AISC *Steel Construction Manual* Table 17-11. For a one foot width of deck, the maximum horizontal reaction, R, is

$$R = \gamma_{TU} E_c \alpha \Delta T t_s$$
$$= (1.00)\left(3152 \ \frac{\text{kips}}{\text{in}^2}\right)\left(6.0 \times 10^{-6} \ \frac{1}{°\text{F}}\right)$$
$$\times (30°\text{F})(8 \text{ in})\left(12 \ \frac{\text{in}}{\text{ft}}\right)$$
$$= 54.5 \text{ kips/ft} \quad (55 \text{ kips/ft})$$

The answer is (C).

4. From ASCE/SEI7 Eq. 10.4-4, the height factor, f_z, is

$$f_z = \left(\frac{z}{33 \text{ ft}}\right)^{0.10} = \left(\frac{20 \text{ ft}}{33 \text{ ft}}\right)^{0.10} = 0.951$$

From ASCE/SEI7 Table 1.5-1, an essential facility is in risk category IV.

From ASCE/SEI7 Table 1.5-2, the ice importance factor—thickness, I_i, for a risk category IV building is 1.25.

The design ice thickness, t_d, is given by ASCE/SEI7 Eq. 10.4-5.

$$t_d = 2.0 t I_i f_z K_{zt}^{0.35}$$
$$= (2.0)(0.75 \text{ in})(1.25)(0.951)(1.0)^{0.35}$$
$$= 1.78 \text{ in}$$

The width, b_f, and depth, d, of a W10 × 12 beam are given in AISC *Steel Construction Manual* Table 1-1. From ASCE/SEI7 Fig. 10-1, the diameter of a cylinder circumscribing a wide flange beam, D_c, is

$$D_c = \sqrt{b_f^2 + d^2} = \sqrt{(3.96 \text{ in})^2 + (9.87 \text{ in})^2}$$
$$= 10.63 \text{ in}$$

From ASCE/SEI7 Eq. 10.4-1, the cross-sectional area of ice is

$$A_i = \pi t_d (D_c + t_d)$$
$$= \pi(1.78 \text{ in})(10.63 \text{ in} + 1.78 \text{ in})$$
$$= 69.4 \text{ in}^2$$

The unfactored ice load, w_i, is the area of ice multiplied by the density.

$$w_i = A_i \gamma_i = \left(\frac{69.4 \text{ in}^2}{\left(12 \ \frac{\text{in}}{\text{ft}}\right)^2}\right)\left(57.2 \ \frac{\text{lbf}}{\text{ft}^3}\right)$$
$$= 27.6 \text{ lbf/ft} \quad (28 \text{ lbf/ft})$$

The answer is (B).

5. Horizontal pressures due to surcharge loads are given in AASHTO Sec. 3.11.6.2. From AASHTO Eq. 3.11.6.2-3, the horizontal pressure due to an infinitely long line load parallel to a wall is

$$\Delta_{ph} = \left(\frac{4Q}{\pi}\right)\left(\frac{X^2 Z}{R^4}\right)$$

From AASHTO Fig. 3.11.6.2-2, the distance, R, is the radial distance between the point of load application and the point on the wall under consideration.

$$R = \sqrt{X^2 + Z^2} = \sqrt{(6 \text{ ft})^2 + (4 \text{ ft})^2} = 7.21 \text{ ft}$$

The horizontal pressure on the wall, Δ_{ph}, is

$$\Delta_{ph} = \left(\frac{4Q}{\pi}\right)\left(\frac{X^2 Z}{R^4}\right) = \left(\frac{(4)\left(640 \; \frac{lbf}{ft}\right)}{\pi}\right)\left(\frac{(6 \; ft)^2 (4 \; ft)}{(7.21 \; ft)^4}\right)$$

$$= 43 \; lbf/ft^2$$

The answer is (D).

6. The design water velocity, v, is equal to the water flow rate divided by the area, A, of the stream at the design flood. (AASHTO uses V for velocity, whereas this solution uses v.)

$$v = \frac{Q}{A} = \frac{6000 \; \frac{ft^3}{sec}}{(8 \; ft)(50 \; ft)} = 15 \; ft/sec$$

For $\theta = 20°$, AASHTO Table 3.7.3.2-1 gives a lateral drag coefficient, C_L, of 0.9.

Use AASHTO Eq. 3.7.3.2-1 to find the lateral water pressure, p. (This equation is not dimensionally consistent.)

$$p = \frac{C_L v^2}{1000} = \frac{(0.9)\left(15 \; \frac{ft}{sec}\right)^2}{1000} = 0.203 \; kips/ft^2$$

The lateral water pressure acts at a distance, d, above the bottom of the pier. The distance is

$$d = 2 \; ft + \frac{h}{2} = 2 \; ft + \frac{8 \; ft}{2} = 6 \; ft$$

Treating the pier as a cantilever, the moment, M, at the base due to the stream pressure is

$$M = pBdh = \left(0.203 \; \frac{kips}{ft^2}\right)(20 \; ft)(6 \; ft)(8 \; ft)$$

$$= 195 \; ft\text{-}kips \quad (190 \; ft\text{-}kips)$$

The answer is (D).

7. Find the weight, W, of the masonry tributary to one expansion anchor. t is the masonry thickness and s is the anchor spacing.

$$W = \gamma_m tsh = \frac{\left(120 \; \frac{lbf}{ft^3}\right)\left(\dfrac{4 \; in}{12 \; \frac{in}{ft}}\right)(2 \; ft)(40 \; ft)}{1000 \; \frac{lbf}{kip}}$$

$$= 3.2 \; kips$$

The eccentricity, e, of the load relative to the face of the masonry wall is

$$e = b_{cavity} + \frac{t}{2} = 2.5 \; in + \frac{4 \; in}{2} = 4.5 \; in$$

The moment, M, due to the eccentricity is

$$M = We = (3.2 \; kips)(4.5 \; in) = 14.4 \; in\text{-}kips$$

The tensile force, T, in the anchor is equal to the moment divided by the distance, d, between the anchor and the centroid of the triangular compression block.

$$T = \frac{M}{d} = \frac{14.4 \; in\text{-}kips}{4.5 \; in - \left(\dfrac{1}{3}\right)(1 \; in)} = 3.5 \; kips$$

The answer is (D).

8. To construct the moment diagram, first draw a free-body diagram of the portion of the beam to the left of the hinge.

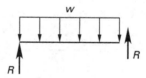

The loaded portion of the beam is simply supported, so the moment diagram will follow the shape of a simple beam (see AISC *Steel Construction Manual* Table 3-23, case 1).

Draw a free-body diagram of the portion of the beam to the right of the hinge.

The only load acting on this portion of the beam is the point load reaction from the simple beam to the left of the hinge. This condition is equivalent to the case of a propped cantilever with a point load at the end of the cantilever (see AISC *Steel Construction Manual* Table 3-23, case 26).

Combining these two moment diagrams yields option A. (This problem uses the convention of positive moments drawn above the beam.)

The problem can also be solved quickly by process of elimination. For a beam with an internal hinge, the moment must be zero at the hinge. Therefore, options B and D are incorrect. Also, discontinuities in the moment diagram can only occur at locations of applied concentrated loads or reactions; therefore, option C is incorrect.

The answer is (A).

9. Option D is the preferred option because the hanger loads are transferred to the wood beam through bearing. This configuration avoids tension perpendicular to grain and stress concentrations at locations of fasteners.

Hanging multiple loads below the neutral axis of a wood beam is not recommended. This configuration causes tension stresses perpendicular to the grain and may cause horizontal splits in the wood. For this reason, options A and B should be avoided.

The fasteners in option C are above the neutral axis, but this can still lead to splitting in the wood. The steel plate with a vertical row of bolts will restrain moisture shrinkage of the wood, resulting in stresses perpendicular to the grain.

The answer is (D).

10. From AASHTO Sec. 4.6.2.2.1, e_g is the distance between the centers of gravity of the beam and the deck.

$$e_g = \frac{t_s + d}{2} = \frac{8 \text{ in} + 38.6 \text{ in}}{2} = 23.3 \text{ in}$$

From AASHTO Eq. 4.6.2.2.1-1, the longitudinal stiffness parameter, K_g, is

$$K_g = n(I + A e_g^2)$$
$$= (8)\left(11{,}600 \text{ in}^4 + \left((49.2 \text{ in}^2)(23.3 \text{ in})^2\right)\right)$$
$$= 306{,}500 \text{ in}^4$$

AASHTO Table 4.6.2.2.2b-1 gives live load distribution factors for moment in interior beams. From AASHTO Table 4.6.2.2.1-1, the deck superstructure cross section is type (a).

From the problem illustration,

$$S = 10 \text{ ft}$$
$$t_s = 8 \text{ in}$$
$$L = 75 \text{ ft}$$
$$N_b = 4$$

All parameters are within the range of applicability from AASHTO Table 4.6.2.2.2b-1.

Find the live load distribution factor for two or more design lanes loaded. (This equation is not dimensionally consistent.)

$$0.075 + \left(\frac{S}{9.5}\right)^{0.6}\left(\frac{S}{L}\right)^{0.2}\left(\frac{K_g}{12.0 L t_s^3}\right)^{0.1}$$
$$= 0.075 + \left(\frac{10 \text{ ft}}{9.5}\right)^{0.6}\left(\frac{10 \text{ ft}}{75 \text{ ft}}\right)^{0.2}$$
$$\times \left(\frac{306{,}500 \text{ in}^4}{(12.0)(75 \text{ ft})(8 \text{ in})^3}\right)^{0.1}$$
$$= 0.737 \text{ lanes} \quad (0.74 \text{ lanes})$$

The answer is (C).

11. Creep under sustained, long-term loading is not a design consideration for steel under normal temperatures. Therefore, option A is incorrect.

P-delta effects do not apply for the given configuration and loading. Therefore, option B is incorrect.

The provided calculation includes an imposed load of 100 kips. The self-weight of the beams is

$$W = w_{\text{W10}\times22}L_1 + w_{\text{W10}\times68}L_2$$
$$= \left(\frac{22 \dfrac{\text{lbf}}{\text{ft}}}{1000 \dfrac{\text{lbf}}{\text{kip}}}\right)(2)(5 \text{ ft}) + \left(\frac{68 \dfrac{\text{lbf}}{\text{ft}}}{1000 \dfrac{\text{lbf}}{\text{kip}}}\right)(8 \text{ ft})$$
$$= 0.764 \text{ kips}$$

This self-weight is negligible compared to the imposed load of 100 kips. Therefore, option C is incorrect.

Therefore, by process of elimination, the calculation done by hand ignores deflection due to shear deformations (and includes only deflection due to flexural deformations).

Check mathematically.

The span-to-depth ratio is small, and shear deformations will contribute to the overall deflection. The deflection due to shear deformations, Δ_{shear}, is found using the equation

$$\Delta_{\text{shear}} = \frac{PL_1}{4A_s G}$$

The shear modulus, G, is found from the modulus of elasticity, E, and Poisson's ratio, ν. For steel, ν is 0.3.

$$G = \frac{E}{2(1+\nu)} = \frac{29{,}000 \ \dfrac{\text{kips}}{\text{in}^2}}{(2)(1+0.3)}$$
$$= 11{,}200 \ \text{kips/in}^2$$

For wide flange shapes, the shear area, A_s, can be approximated as the area of the web. Beam properties for a W10 × 22 are found in AISC *Steel Construction Manual* Table 1-1.

$$A_s = (d - 2t_f)t_w = \big(10.2 \ \text{in} - (2)(0.36 \ \text{in})\big)(0.24 \ \text{in})$$
$$= 2.275 \ \text{in}^2$$

The deflection due to shear deformations is

$$\Delta_{\text{shear}} = \frac{PL_1}{4A_s G} = \frac{\left(\dfrac{100 \ \text{kips}}{2}\right)(5 \ \text{ft})\left(12 \ \dfrac{\text{in}}{\text{ft}}\right)}{(4)(2.275 \ \text{in}^2)\left(11{,}200 \ \dfrac{\text{kips}}{\text{in}^2}\right)}$$
$$= 0.0294 \ \text{in}$$

Therefore, shear deformations account for the discrepancy of

$$0.096 \ \text{in} - 0.066 \ \text{in} = 0.030 \ \text{in} \approx 0.0294 \ \text{in}$$

The answer is (D).

12. The beam is statically indeterminate to the first degree, as there are three equations of equilibrium and four unknown reactions. One method of solving for the interior support reaction is to use deflection compatibility. Start by removing the interior support to create a statically determinant beam, and then find the deflection at the removed support.

Δ_{max} is given by AISC *Steel Construction Manual* Table 3-23, case 9 (when $a = l/3$). P is 20 kips, and L is 60 ft.

$$\Delta_{\text{max}} = \frac{PL^3}{28EI}$$

Apply an upward reaction, R, at the removed support, and then find the deflection at the removed support as a function of R.

From AISC *Steel Construction Manual* Table 3-23, case 7, taking $P = R$,

$$\Delta_{\text{max}} = \frac{RL^3}{48EI}$$

The total deflection at the middle support must be zero, so set the two deflections equal to each other and solve for the reaction, R, at the interior support.

$$\frac{PL^3}{28EI} = \frac{RL^3}{48EI}$$
$$R = \frac{48P}{28} = \frac{(48)(20 \ \text{kips})}{28} = 34 \ \text{kips}$$

The answer is (C).

13. To construct the influence line for a truss, apply a unit load to each of the joints of the bottom chord and calculate the corresponding force in member CG. Then, plot the points and connect the values with straight lines.

From a free body diagram of joint C, member CG is the only member connected to the joint with a vertical component. Therefore, for the case of a unit load applied to joints A, B, D, and E, there is no vertical force acting on joint C, and member CG must be a zero-force member. For a unit load applied to joint C, member CG must resist the full applied load. Therefore, member CG has an axial load of 1 (tension). Use straight lines to connect the joint values of A = 0, B = 0, C = 1, D = 0, and E = 0.

The influence diagram is

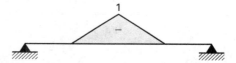

The answer is (A).

14. ASCE/SEI7 Table 4-1 gives the minimum uniformly distributed live loads for different occupancies and uses. For light storage warehouses,

$$L_{o,1} = 125 \ \text{lbf/ft}^2$$

For ordinary flat roofs,

$$L_{o,\text{roof}} = 20 \ \text{lbf/ft}^2$$

From ASCE/SEI7 Sec. 4.7.3 and Table 4-1 footnote a, live loads that exceed 100 lbf/ft² cannot be reduced. The exception in ASCE/SEI7 Sec. 4.7.3 does not apply because the column only supports one floor of storage.

Therefore, the full live load of 125 lbf/ft^2 must be considered at the first floor.

$$L = L_{o,1} = 125 \text{ lbf/ft}^2$$

From ASCE/SEI7 Sec. 4.8.1, the live load may be reduced for ordinary flat roofs. R_1 is a function of the tributary area of a member. From the problem illustration, the tributary area, A_t, of column B-2 is

$$A_t = bl = (20 \text{ ft})(20 \text{ ft}) = 400 \text{ ft}^2$$

Calculate R_1 from ASCE/SEI7 Sec. 4.8.2, for 200 ft$^2 <$ $A_t < 600$ ft^2. (This equation is not dimensionally consistent.)

$$R_1 = 1.2 - 0.001A_t = 1.2 - (0.001)(400 \text{ ft}^2) = 0.8$$

R_2 is 1.0 for flat roofs.

From ASCE/SEI7 Eq. 4.8-1, the reduced roof live load, L_r, is

$$L_r = L_{o,\text{roof}}R_1R_2 = \left(20\,\frac{\text{lbf}}{\text{ft}^2}\right)(0.8)(1.0) = 16 \text{ lbf/ft}^2$$

The design axial load, P_u, is

$$P_u = A_t(1.2D + 1.6L_r + L)$$

$$= \frac{(400 \text{ ft}^2)\begin{pmatrix}(1.2)\left(40\,\dfrac{\text{lbf}}{\text{ft}^2} + 60\,\dfrac{\text{lbf}}{\text{ft}^2}\right)\\[2mm] +(1.6)\left(16\,\dfrac{\text{lbf}}{\text{ft}^2}\right) + (1.0)\left(125\,\dfrac{\text{lbf}}{\text{ft}^2}\right)\end{pmatrix}}{1000\,\dfrac{\text{lbf}}{\text{kip}}}$$

$$= 108 \text{ kips}$$

The answer is (C).

15. Steel-framed floor systems are susceptible to floor vibrations and can increase occupant discomfort. Floors with lower natural frequencies are most susceptible to these vibrations. The natural frequency of a floor will decrease as the spans get larger and the depths (i.e., moment of inertia) of beams get smaller. Therefore, options B and C are incorrect.

Additionally, steel-framed floors that are intended for uses that involve rhythmic activities, such as aerobics and dancing, must be designed to meet vibration serviceability requirements. A building originally designed for office occupancy will not meet the vibration requirements of a gym and will be susceptible to floor vibrations. Therefore, option D is incorrect.

Concrete floors are less susceptible to vibration problems than steel-framed floors. Meeting the code-prescribed slab thickness and deflection limits is generally sufficient to satisfy vibration serviceability requirements for concrete floors.

(Refer to AISC *Design Guide 11: Floor Vibrations Due to Human Activity* for further information regarding vibration design.)

The answer is (A).

16. For the fatigue II load combination and a finite bridge life of 75 years, the nominal fatigue resistance, $(\Delta F)_n$, is given by AASHTO Eq. 6.6.1.2.5-2 as

$$(\Delta F)_n = \left(\frac{A}{N}\right)^{1/3}$$

From AASHTO Eq. 6.6.1.2.5-3,

$$N = (365)(75)n(\text{ADTT})_{\text{SL}}$$

For simply supported girders with a span length greater than 40 ft, AASHTO Table 6.6.1.2.5-2 gives the number of stress range cycles per truck passage, n, as 1.0 cycle/truck.

From AASHTO Sec. 6.6.1.2.5, the single-lane average daily truck traffic, ADTT, is specified in AASHTO Sec. 3.6.1.4. From AASHTO Eq. 3.6.1.4.2-1,

$$(\text{ADTT})_{\text{SL}} = p(\text{ADTT})$$

From AASHTO Table 3.6.1.4.2-1, for three or more lanes available to trucks, $p = 0.80$.

So,

$$(\text{ADTT})_{\text{SL}} = p(\text{ADTT}) = (0.80)\left(1000\,\frac{\text{trucks}}{\text{day}}\right)$$
$$= 800 \text{ trucks/day}$$

The number of cycles of stress range, N, is

$$N = \left(365\,\frac{\text{days}}{\text{yr}}\right)(\text{bridge life})n(\text{ADTT})_{\text{SL}}$$
$$= \left(365\,\frac{\text{days}}{\text{yr}}\right)(75 \text{ yr})\left(1.0\,\frac{\text{cycles}}{\text{truck}}\right)\left(800\,\frac{\text{trucks}}{\text{day}}\right)$$
$$= 2.19 \times 10^7 \text{ cycles}$$

The problem illustration matches AASHTO Table 6.6.1.2.3-1, description 4.1, which gives a detail category of C'. From that table, or from AASHTO Table 6.6.1.2.5-1, the detail category constant, A, is

$$A = 44.0 \times 10^8 \left(\frac{\text{kips}}{\text{in}^2} \right)^3$$

From AASHTO Eq. 6.6.1.2.5-2, the nominal fatigue resistance for the welded stiffener connection, $(\Delta F)_n$, is

$$(\Delta F)_n = \left(\frac{A}{N} \right)^{1/3}$$

$$= \left(\frac{44.0 \times 10^8 \left(\frac{\text{kips}}{\text{in}^2} \right)^3}{2.19 \times 10^7} \right)^{1/3}$$

$$= 5.9 \text{ kips/in}^2$$

The answer is (B).

17. ASD Solution

From AISC *Steel Construction Manual* Chap. 14 Eq. 14-7b, the minimum base plate thickness, $t_{\min}$, is

$$t_{\min} = l \sqrt{\frac{3.33 P_a}{F_y BN}}$$

l is the largest of m, n, and $\lambda n'$.

From AISC *Steel Construction Manual* Table 1-1, the relevant section properties for a W14 × 109 column are

$$d = 14.3 \text{ in}$$
$$b_f = 14.6 \text{ in}$$

The base plate is designed for the full available axial strength of the column. So, from AISC *Steel Construction Manual* Table 4-1, the axial strength of a W14 × 109 with an effective length of 15 ft is

$$P_a = \frac{P_n}{\Omega_c} = 808 \text{ kips}$$

From AISC *Steel Construction Manual* Chap. 14 Eq. 14-2, the base plate cantilever dimension parallel to the column web, m, is

$$m = \frac{N - 0.95d}{2} = \frac{18 \text{ in} - (0.95)(14.3 \text{ in})}{2}$$
$$= 2.21 \text{ in}$$

From Eq. 14-3, the base plate cantilever dimension parallel to the column flange, n, is

$$n = \frac{B - 0.8 b_f}{2} = \frac{18 \text{ in} - (0.8)(14.6 \text{ in})}{2}$$
$$= 3.16 \text{ in}$$

Assuming $\lambda = 1.0$ per the problem statement, the modified base plate cantilever dimension parallel to the column flange, $\lambda n'$, is

$$\lambda n' = \lambda \left(\frac{\sqrt{d b_f}}{4} \right) = (1.0) \left(\frac{\sqrt{(14.3 \text{ in})(14.6 \text{ in})}}{4} \right)$$
$$= 3.61 \text{ in} \quad [\text{controls}]$$

Therefore, $l = 3.61$ in.

The minimum base plate thickness is

$$t_{\min} = l \sqrt{\frac{3.33 P_a}{F_y BN}} = (3.61 \text{ in}) \sqrt{\frac{(3.33)(808 \text{ kips})}{\left(36 \frac{\text{kips}}{\text{in}^2} \right)(18 \text{ in})(18 \text{ in})}}$$
$$= 1.73 \text{ in} \quad (1\tfrac{3}{4} \text{ in})$$

The answer is (D).

LRFD Solution

From AISC *Steel Construction Manual* Chap. 14 Eq. 14-7a, the minimum thickness, $t_{\min}$, is

$$t_{\min} = l \sqrt{\frac{2 P_u}{0.9 F_y BN}}$$

l is the largest of m, n, and $\lambda n'$.

From AISC *Steel Construction Manual* Table 1-1, the relevant section properties for a W14 × 109 column are

$$d = 14.3 \text{ in}$$
$$b_f = 14.6 \text{ in}$$

The base plate is designed for the full available axial strength of the column. So, from AISC *Steel Construction Manual* Table 4-1, the axial strength of a W14 × 109 with an effective length of 15 ft is

$$P_u = \phi_c P_n = 1210 \text{ kips}$$

From AISC *Steel Construction Manual* Chap. 14 Eq. 14-2, the base plate cantilever dimension parallel to the column web, m, is

$$m = \frac{N - 0.95d}{2} = \frac{18 \text{ in} - (0.95)(14.3 \text{ in})}{2}$$
$$= 2.21 \text{ in}$$

From Eq. 14-3, the base plate cantilever dimension parallel to the column flange, n, is

$$n = \frac{B - 0.8b_f}{2} = \frac{18 \text{ in} - (0.8)(14.6 \text{ in})}{2}$$
$$= 3.16 \text{ in}$$

Assuming $\lambda = 1.0$ per the problem statement, the modified base plate cantilever dimension parallel to the column flange, $\lambda n'$, is

$$\lambda n' = \lambda \left(\frac{\sqrt{db_f}}{4} \right) = (1.0) \left(\frac{\sqrt{(14.3 \text{ in})(14.6 \text{ in})}}{4} \right)$$
$$= 3.61 \text{ in} \quad [\text{controls}]$$

Therefore, $l = 3.61$ in.

The minimum base plate thickness is

$$t_{\min} = l \sqrt{\frac{2P_u}{0.9F_y BN}}$$
$$= (3.61 \text{ in}) \sqrt{\frac{(2)(1210 \text{ kips})}{(0.9)\left(36 \dfrac{\text{kips}}{\text{in}^2}\right)(18 \text{ in})(18 \text{ in})}}$$
$$= 1.73 \text{ in} \quad (1\tfrac{3}{4} \text{ in})$$

The answer is (D).

18. Find the maximum bending stress in the existing beam prior to the installation of the reinforcement, f_{b1}.

$$f_{b1} = \frac{M_1}{S_x} = \frac{(20 \text{ ft-kips})\left(12 \dfrac{\text{in}}{\text{ft}}\right)}{34.1 \text{ in}^3} = 7.0 \text{ kips/in}^2$$

Find the maximum bending stress at the top of the composite section, f_{b2}, due to the additional moment, M_2.

From AISC *Steel Construction Manual* Table 1-8, the properties of a WT8 × 25 shape are

$$A_{\text{WT}} = 7.37 \text{ in}^2$$
$$d_{\text{WT}} = 8.13 \text{ in}$$
$$I_{x,\text{WT}} = 42.3 \text{ in}^4$$
$$\bar{y}_{\text{WT}} = 1.89 \text{ in}$$

Measured from the bottom of the WT8 × 25, the centroid of the composite reinforced section, $\bar{y}_{\text{comp}}$, is

$$\bar{y}_{\text{comp}} = \frac{A_{\text{WT}}\bar{y}_{\text{WT}} + A\left(d_{\text{WT}} + \dfrac{d}{2}\right)}{A_{\text{WT}} + A}$$

$$= \frac{\begin{array}{c}(7.37 \text{ in}^2)(1.89 \text{ in}) + (7.97 \text{ in}^2) \\ \times \left(8.13 \text{ in} + \dfrac{11.96 \text{ in}}{2}\right)\end{array}}{7.37 \text{ in}^2 + 7.97 \text{ in}^2}$$

$$= 8.24 \text{ in}$$

The moment of inertia of the composite reinforced section, I_{comp}, is

$$I_{\text{comp}} = I_x + A\left(\left(d_{\text{WT}} + \frac{d}{2}\right) - \bar{y}_{\text{comp}}\right)^2$$
$$\quad + I_{x,\text{WT}} + A_{\text{WT}}(\bar{y}_{\text{comp}} - \bar{y}_{\text{WT}})^2$$
$$= 204.1 \text{ in}^4 + (7.97 \text{ in}^2)$$
$$\quad \times \left(\left(8.13 \text{ in} + \frac{11.96 \text{ in}}{2}\right) - 8.24 \text{ in}\right)^2$$
$$\quad + 42.3 \text{ in}^4 + (7.37 \text{ in}^2)(8.24 \text{ in} - 1.89 \text{ in})^2$$
$$= 818.2 \text{ in}^4$$

The distance from the centroid of the reinforced section to the top fiber, c_{top}, is

$$c_{\text{top}} = (d_{\text{WT}} + d) - \bar{y}_{\text{comp}} = (8.13 \text{ in} + 11.96 \text{ in}) - 8.24 \text{ in}$$
$$= 11.85 \text{ in}$$

The maximum bending stress at the top of the existing beam due to the additional moment is

$$f_{b2} = \frac{M_2 c_{\text{top}}}{I_{\text{comp}}} = \frac{(80 \text{ ft-kips})(11.85 \text{ in})\left(12 \dfrac{\text{in}}{\text{ft}}\right)}{818.2 \text{ in}^4}$$
$$= 13.9 \text{ kips/in}^2$$

The total maximum bending stress, $f_{b,\text{tot}}$, is

$$f_{b,\text{tot}} = f_{b1} + f_{b2} = 7.0 \ \frac{\text{kips}}{\text{in}^2} + 13.9 \ \frac{\text{kips}}{\text{in}^2}$$
$$= 20.9 \ \text{kips/in}^2 \quad (21 \ \text{kips/in}^2)$$

The answer is (C).

19. ASD Solution

The maximum moment demand, M_r, is given in AISC *Steel Construction Manual* Table 3-23, case 7 as

$$M_r = \frac{P_a L}{4} = \frac{(10 \ \text{kips} + 29 \ \text{kips})(12 \ \text{ft})}{4} = 117 \ \text{ft-kips}$$

The moment due to the self-weight of the beam is negligible.

From AISC *Steel Construction Manual* Table 3-1, for a simply supported beam with a concentrated load at midspan, the lateral-torsional buckling modification factor, C_b, is 1.32.

AISC *Steel Construction Manual* Table 3-10 gives available moments versus unbraced lengths for $C_b = 1.0$. Adjust the moment demand by the determined value of C_b, and enter the table on the vertical axis at

$$\frac{M_r}{C_b} = \frac{117 \ \text{ft-kips}}{1.32} = 88.6 \ \text{ft-kips}$$

On the horizontal axis, enter the table at $L_b = 12$ ft.

Move up the vertical axis until the first solid line is reached at $M_n/\Omega = 103$ ft-kips.

Select a W14 × 34.

From AISC 360 Eq. F2-2, verify the flexural strength and check that M_p does not control.

From AISC *Steel Construction Manual* Table 3-2, for a W14 × 34 shape,

$$\frac{M_p}{\Omega} = 136 \ \text{ft-kips}$$

From AISC 360 Eq. F2-2,

$$\frac{M_n}{\Omega} = \frac{C_b\left[M_p - (M_p - 0.7F_y S_x)\left(\dfrac{L_b - L_p}{L_r - L_p}\right)\right]}{\Omega} \leq \frac{M_p}{\Omega}$$
$$= (1.32)(103 \ \text{ft-kips})$$
$$= 136 \ \text{ft-kips} \leq 136 \ \text{ft-kips} \quad [M_p \ \text{does not control}]$$

Because 136 ft-kips is greater than 117 ft-kips, the shape is adequate.

Check that shear, V_r, does not control.

$$V_r = \frac{P_a}{2} = \frac{10 \ \text{kips} + 29 \ \text{kips}}{2} = 19.5 \ \text{kips}$$

From AISC *Steel Construction Manual* Table 3-2, for a W14 × 34 shape,

$$\frac{V_{nx}}{\Omega_v} = 79.8 \ \text{kips} > 19.5 \ \text{kips} \quad [\text{shear does not control}]$$

Therefore, a W14 × 34 is the lightest sufficient shape.

The answer is (B).

LRFD Solution

For LRFD, the controlling load combination will be $1.2D + 1.6L$. The maximum moment demand, M_r, is given in AISC *Steel Construction Manual* Table 3-23, case 7 as

$$M_r = \frac{P_u L}{4} = \frac{\big((1.2)(10 \ \text{kips}) + (1.6)(29 \ \text{kips})\big)(12 \ \text{ft})}{4}$$
$$= 175 \ \text{ft-kips}$$

The moment due to the self-weight of the beam is negligible.

From AISC *Steel Construction Manual* Table 3-1, for a simply supported beam with a concentrated load at midspan, the lateral-torsional buckling modification factor, C_b, is 1.32.

AISC *Steel Construction Manual* Table 3-10 gives available moments versus unbraced lengths for $C_b = 1.0$. Adjust the moment demand by the determined value of C_b, and enter the table on the vertical axis at

$$\frac{M_r}{C_b} = \frac{175 \ \text{ft-kips}}{1.32} = 133 \ \text{ft-kips}$$

On the horizontal axis, enter the table at $L_b = 12$ ft.

Move up the vertical axis until the first solid line is reached at $\phi M_n = 154.5$ ft-kips.

Select a W14 × 34.

From AISC 360 Eq. F2-2, verify the flexural strength, and check that M_p does not control.

From AISC *Steel Construction Manual* Table 3-2, for a W14 × 34 shape,

$$\phi M_p = 205 \ \text{ft-kips}$$

From AISC 360 Eq. F2-2,

$$\phi M_n = \phi C_b \left(M_p - (M_p - 0.7F_y S_x) \left(\frac{L_b - L_p}{L_r - L_p} \right) \right) \leq \phi M_p$$

$$= (1.32)(154.5 \text{ ft-kips})$$

$$= 204 \text{ ft-kips} \leq 205 \text{ ft-kips} \quad [M_p \text{ does not control}]$$

Because 205 ft-kips is greater than 175 ft-kips, the shape is adequate.

Check that shear, V_r, does not control.

$$V_r = \frac{P_u}{2} = \frac{(1.2)(10 \text{ kips}) + (1.6)(29 \text{ kips})}{2} = 29.2 \text{ kips}$$

From AISC *Steel Construction Manual* Table 3-2, for a W14 × 34 shape,

$$\phi V_{nx} = 120 \text{ kips} > 29.2 \text{ kips} \quad [\text{shear does not control}]$$

Therefore, a W14 × 34 is the lightest sufficient shape.

The answer is (B).

20. **ASD Solution**

From AISC *Steel Construction Manual* Table 8-8, the minimum weld size (in sixteenths-of-an-inch), D_{min}, is given by the following equation. (This equation is not dimensionally consistent.)

$$D_{min} = \frac{\Omega P_a}{CC_1 l}$$

Determine C using AISC *Steel Construction Manual* Table 8-8. The eccentricity, e_x, of the load is determined from the problem illustration as

$$e_x = 6 \text{ in} + 3 \text{ in} - 0.62 \text{ in} = 8.38 \text{ in}$$

Find a and k.

$$a = \frac{e_x}{l} = \frac{8.38 \text{ in}}{8.5 \text{ in}} = 0.99$$

$$k = \frac{b}{l} = \frac{3 \text{ in}}{8.5 \text{ in}} = 0.35$$

Interpolating from AISC *Steel Construction Manual* Table 8-8, $C = 1.34$.

Determine the minimum weld size. Ω is 2.00 (from *AISC Manual* Table 8-8), C_1 is 1.0 (from *AISC Manual* Table 8-3), l is given as 8.5 in, and P_a is given as 16 kips.

$$D_{min} = \frac{\Omega P_a}{CC_1 l} = \frac{(2.00)(16 \text{ kips})}{(1.34)(1.0)(8.5 \text{ in})} = 2.81 \text{ sixteenths}$$

Therefore, use a $\frac{3}{16}$ in fillet weld.

Check the minimum weld size requirements. From AISC 360 Table J2.4, a $\frac{3}{16}$ in fillet weld is acceptable for materials that are between $\frac{1}{4}$ in and $\frac{1}{2}$ in thick.

From AISC *Steel Construction Manual* Table 1-8, for a WT9 × 43,

$$t_w = 0.480 \text{ in} \quad [\text{OK}]$$

From AISC *Steel Construction Manual* Table 1-1, for a W12 × 40,

$$t_w = 0.295 \text{ in} \quad [\text{OK}]$$

The answer is (A).

LRFD Solution

From AISC *Steel Construction Manual* Table 8-8, the minimum weld size (in sixteenths-of-an-inch), D_{min}, is given by the following equation. (The equation is not dimensionally consistent.)

$$D_{min} = \frac{P_u}{\phi CC_1 l}$$

Determine C using AISC *Steel Construction Manual* Table 8-8. The eccentricity, e_x, of the load is determined from the problem illustration as

$$e_x = 6 \text{ in} + 3 \text{ in} - 0.62 \text{ in} = 8.38 \text{ in}$$

Find a and k.

$$a = \frac{e_x}{l} = \frac{8.38 \text{ in}}{8.5 \text{ in}} = 0.99$$

$$k = \frac{b}{l} = \frac{3 \text{ in}}{8.5 \text{ in}} = 0.35$$

Interpolating from AISC *Steel Construction Manual* Table 8-8, $C = 1.34$.

Determine the minimum weld size. ϕ is 0.75 (from AISC *Steel Construction Manual* Table 8-8), C_1 is 1.0 (from AISC *Steel Construction Manual* Table 8-3), l is given as 8.5 in, and P_u is given as 24 kips.

$$D_{min} = \frac{P_u}{\phi CC_1 l}$$

$$= \frac{24 \text{ kips}}{(0.75)(1.34)(1.0)(8.5 \text{ in})}$$

$$= 2.81 \text{ sixteenths}$$

Therefore, use a $\frac{3}{16}$ in fillet weld.

Check the minimum weld size requirements. From AISC 360 Table J2.4, a $^3/_{16}$ in fillet weld is acceptable for materials that are between $^1/_4$ in and $^1/_2$ in thick.

From AISC *Steel Construction Manual* Table 1-8, for a WT9 × 43,

$$t_w = 0.480 \text{ in} \quad [\text{OK}]$$

From AISC *Steel Construction Manual* Table 1-1, for a W12 × 40,

$$t_w = 0.295 \text{ in} \quad [\text{OK}]$$

The answer is (A).

21. AASHTO App. D6, Table D6.1-1 gives equations for the plastic moment of the composite section, M_p, based on different locations of the plastic neutral axis. Because $\overline{Y} < t_s$, the plastic neutral axis is within the concrete deck. Therefore, cases III, IV, V, VI, and VII in AASHTO App. D6, Table D6.1-1 are applicable. Ignoring the contribution from the slab reinforcement means that $P_{rt} = P_{rb} = 0$ kips, so cases III through VII all simplify to the same equation for M_p.

$$M_p = \frac{\overline{Y}^2 P_s}{2t_s} + (P_c d_c + P_w d_w + P_t d_t)$$

Though $\overline{Y}$ was given in the problem statement, it can also be found by equating the steel force with the concrete force.

$$\overline{Y} = \frac{F_y A_s}{0.85 f'_c b_s} = \frac{\left(50 \frac{\text{kips}}{\text{in}^2}\right)(53.3 \text{ in}^2)}{(0.85)\left(4 \frac{\text{kips}}{\text{in}^2}\right)(120 \text{ in})} = 6.53 \text{ in}$$

Because the steel section is a symmetrical wide flange shape, the moment due to the forces in the top flange, web, and bottom flange of the steel beam, $P_c d_c + P_w d_w + P_t d_t$, is equal to the area of the steel beam, multiplied by the yield stress of the beam, multiplied by the distance between the centroid of the beam and the plastic neutral axis of the composite section.

$$
\begin{aligned}
P_c d_c &+ P_w d_w + P_t d_t \\
&= F_y A_s \left(\frac{d}{2} + 2 \text{ in} + t_s - \overline{Y}\right) \\
&= \left(50 \frac{\text{kips}}{\text{in}^2}\right)(53.3 \text{ in}^2) \\
&\quad \times \left(\frac{39 \text{ in}}{2} + 2 \text{ in} + 8 \text{ in} - 6.53 \text{ in}\right) \\
&= 61{,}215 \text{ in-kips}
\end{aligned}
$$

From AASHTO Sec. 4.6.2.6, the effective flange width, b_s, for an interior girder is equal to the girder spacing. Therefore, b_s is

$$b_s = (10 \text{ ft})\left(12 \frac{\text{in}}{\text{ft}}\right) = 120 \text{ in}$$

The axial force in the slab, P_s, is

$$
\begin{aligned}
P_s &= 0.85 f'_c b_s t_s \\
&= (0.85)\left(4 \frac{\text{kips}}{\text{in}^2}\right)(120 \text{ in})(8 \text{ in}) \\
&= 3264 \text{ kips}
\end{aligned}
$$

The plastic moment of the composite section is

$$
\begin{aligned}
M_p &= \frac{\overline{Y}^2 P_s}{2t_s} + (P_c d_c + P_w d_w + P_t d_t) \\
&= \frac{\dfrac{(6.53 \text{ in})^2 (3264 \text{ kips})}{(2)(8 \text{ in})} + 61{,}215 \text{ in-kips}}{12 \frac{\text{in}}{\text{ft}}} \\
&= 5826 \text{ ft-kips}
\end{aligned}
$$

From AASHTO Sec. 6.10.7.1.2, the ratio of D_p to D_t controls which equation will be used for the nominal flexural resistance, M_n.

The total depth of the composite section, D_t, is

$$D_t = d + 2 \text{ in} + t_s = 39 \text{ in} + 2 \text{ in} + 8 \text{ in} = 49 \text{ in}$$

The distance from the top of the concrete deck to the neutral axis of the composite section, D_p, is

$$D_p = \overline{Y} = 6.53 \text{ in}$$

The ratio is

$$\frac{D_p}{D_t} = \frac{6.53 \text{ in}}{49 \text{ in}} = 0.133$$

$D_p/D_t > 0.1$, so use AASHTO Eq. 6.10.7.1.2-2. The

nominal flexural resistance of the composite interior girder is

$$M_n = M_p\left(1.07 - 0.7\left(\frac{D_p}{D_t}\right)\right)$$
$$= (5826 \text{ ft-kips})\left(1.07 - (0.7)(0.133)\right)$$
$$= 5690 \text{ ft-kips}$$

The answer is (C).

22. Web crippling strength values are given in *AISI Manual* Table II-14. A one-flange reaction corresponds with case A in the table. Thus, the nominal web crippling strength is $P_n = 570$ lbf.

Alternative Solution

The depth of the flat portion of the web, h, is

$$h = D - 2R - 2t$$
$$= 6 \text{ in} - (2)(0.0712 \text{ in}) - (2)(0.0451 \text{ in})$$
$$= 5.767 \text{ in}$$

From *AISI Specification* Table C3.4.1-2, $C = 4$, $C_R = 0.14$, $C_N = 0.35$, and $C_h = 0.02$.

From *AISI Specification* Table C3.4.1-2, ftn. 1,

$$\frac{h}{t} = \frac{5.767 \text{ in}}{0.0451 \text{ in}} = 128 \le 200 \quad [\text{OK}]$$
$$\frac{N}{t} = \frac{2 \text{ in}}{0.0451 \text{ in}} = 44 \le 210 \quad [\text{OK}]$$
$$\frac{N}{h} = \frac{2 \text{ in}}{5.767 \text{ in}} = 0.35 \le 2.0 \quad [\text{OK}]$$

From *AISI Specification* Eq. C3.4.1-1, the nominal web crippling strength, P_n, is

$$P_n = Ct^2 F_y \sin\theta\left(1 - C_R\sqrt{\frac{R}{t}}\right)\left(1 + C_N\sqrt{\frac{N}{t}}\right)$$
$$\times\left(1 - C_h\sqrt{\frac{h}{t}}\right)$$
$$= (4)(0.0451 \text{ in})^2\left(33 \frac{\text{kips}}{\text{in}^2}\right)\left(1000 \frac{\text{lbf}}{\text{kip}}\right)\sin 90°$$
$$\times\left(1 - (0.14)\sqrt{\frac{0.0712 \text{ in}}{0.0451 \text{ in}}}\right)$$
$$\times\left(1 + (0.35)\sqrt{\frac{2 \text{ in}}{0.0451 \text{ in}}}\right)$$
$$\times\left(1 - (0.02)\sqrt{\frac{5.767 \text{ in}}{0.0451 \text{ in}}}\right)$$
$$= 570 \text{ lbf}$$

The answer is (C).

23. AASHTO App. A4, Table A4-1 gives deck live load moments per foot width. The tabulated moments already include dynamic load allowance and multipresence factors.

From AASHTO Table 4.6.2.2.1-1, the deck superstructure cross section is type (k).

From AASHTO Sec. 4.6.2.1.6, for a type (k) cross section, the design section for negative moment may be taken at one-third the flange width from the centerline of the girder, not exceeding 15 in.

From the problem illustration, the flange width, b_f, is 2 ft 6 in, or 30 in. Therefore, one-third of the flange width is

$$\frac{b_f}{3} = \frac{30 \text{ in}}{3} = 10 \text{ in} < 15 \text{ in}$$

The distance from the centerline of the girder to the design section for negative moment is 10 in. From the problem illustration, the beam spacing is 14 ft.

Interpolating from AASHTO Table A4-1, the negative deck live load moment is

$$M = 9.3 \text{ ft-kips/ft}$$

The answer is (A).

24. From AASHTO Sec. 9.7.1.1, concrete bridge decks should not be less than 7 in deep.

The answer is (A).

25. From ACI 318 Sec. 24.2.3.4 and Sec. 24.2.3.5, the immediate deflection, Δ_I, is computed with the modulus of elasticity, E_c, and the effective moment of inertia, I_e. I_e is given as 600 in^4/ft.

From ACI 318 Sec. 19.2.2.1(a), E_c is

$$E_c = w_c^{1.5} 33 \sqrt{f_c'} = \left(110 \ \frac{\text{lbf}}{\text{ft}^3}\right)^{1.5} (33) \sqrt{3500 \ \frac{\text{lbf}}{\text{in}^2}}$$
$$= 2{,}252{,}000 \ \text{lbf/in}^2$$

The immediate deflection due to the 100 lbf/ft^2 service load is

$$\Delta_i = \frac{5wL^4}{384 E_c I_e}$$

$$= \frac{(5)\left(\dfrac{100 \ \dfrac{\text{lbf}}{\text{ft}^2}}{12 \ \dfrac{\text{in}}{\text{ft}}}\right)\left((16 \ \text{ft})\left(12 \ \dfrac{\text{in}}{\text{ft}}\right)\right)^4}{(384)\left(2{,}252{,}000 \ \dfrac{\text{lbf}}{\text{in}^2}\right)\left(600 \ \dfrac{\text{in}^4}{\text{ft}}\right)}$$

$$= 0.109 \ \text{in}$$

From ACI 318 Sec. 24.2.4.1.1, additional long-term deflections are computed by multiplying the immediate deflection by the factor λ_Δ.

From ACI 318 Eq. 24.2.4.1.1, the multiplier for additional long term deflection, λ_Δ, is

$$\lambda_\Delta = \frac{\xi}{1 + 50\rho'}$$

Use ACI 318 Sec. 24.2.4.1.2 to find the compression reinforcement ratio, ρ', for no. 5 top bars at 12 in o.c. (0.31 in^2 nominal area per ACI 318 App. A).

$$\rho' = \frac{A_s}{bd} = \frac{0.31 \ \text{in}^2}{(12 \ \text{in})(8.5 \ \text{in})} = 0.00304$$

From ACI 318 Table 24.2.4.1.3, for loads sustained for five or more years, $\xi = 2.0$.

λ_Δ is

$$\lambda_\Delta = \frac{\xi}{1 + 50\rho'} = \frac{2.0}{1 + (50)(0.00304)} = 1.74$$

Therefore, the total midspan deflection due to the sustained load, Δ_{tot}, is

$$\Delta_{\text{tot}} = \Delta_i + \lambda_\Delta \Delta_i = 0.109 \ \text{in} + (1.74)(0.109 \ \text{in})$$
$$= 0.30 \ \text{in}$$

The answer is (D).

26. From the illustration, the effective depth, d, of the beam is

$$d = 8 \ \text{ft} - \frac{6 \ \text{in}}{12 \ \dfrac{\text{in}}{\text{ft}}} = 7.5 \ \text{ft}$$

The vertical reaction at node B, R_B, is

$$R_B = P_u\left(\frac{a}{L}\right) = (400 \ \text{kips})\left(\frac{6 \ \text{ft}}{12 \ \text{ft} + 6 \ \text{ft}}\right) = 133 \ \text{kips}$$

The tensile force, F_u, in the tie is

$$F_u = R_B\left(\frac{a}{d}\right) = (133 \ \text{kips})\left(\frac{12 \ \text{ft}}{7.5 \ \text{ft}}\right) = 213 \ \text{kips}$$

From ACI 318 Table 21.2.1, for strut-and-tie models, the strength reduction factor, ϕ, is 0.75.

From ACI 318 Sec. 23.3.1(b) and Sec. 23.7.2, the required area of flexural reinforcement in the tie, A_{ts}, is

$$A_{ts} = \frac{F_u}{\phi f_y} = \frac{213 \ \text{kips}}{(0.75)\left(60 \ \dfrac{\text{kips}}{\text{in}^2}\right)} = 4.73 \ \text{in}^2 \quad (4.8 \ \text{in}^2)$$

From ACI 318 Sec. 9.9.3.2, the minimum area of flexural tension reinforcement for deep beams must meet the requirements of ACI 318 Sec. 9.6.1. From ACI 318 Sec. 9.6.1.2, the minimum area of flexural reinforcement, $A_{s,\text{min}}$, is

$$A_{s,\text{min}} = \left(\frac{3\sqrt{f_c'}}{f_y}\right) b_w d$$

$$= \left(\frac{(3)\sqrt{5000 \ \dfrac{\text{lbf}}{\text{in}^2}}}{\left(60 \ \dfrac{\text{kips}}{\text{in}^2}\right)\left(1000 \ \dfrac{\text{lbf}}{\text{kip}}\right)}\right)$$

$$\times (12 \ \text{in})\left((7.5 \ \text{ft})\left(12 \ \dfrac{\text{in}}{\text{ft}}\right)\right)$$

$$= 3.82 \ \text{in}^2 < 4.73 \ \text{in}^2 \quad [\text{does not control}]$$

However, the area of flexural reinforcement cannot be less than

$$A_{s,\min} = \frac{200b_w d}{f_y} = \frac{(200)(12 \text{ in})\left((7.5 \text{ ft})\left(12\,\frac{\text{in}}{\text{ft}}\right)\right)}{\left(60\,\frac{\text{kips}}{\text{in}^2}\right)\left(1000\,\frac{\text{lbf}}{\text{kip}}\right)}$$

$$= 3.60 \text{ in}^2 < 4.73 \text{ in}^2 \quad [\text{does not control}]$$

Therefore, the required area of flexural steel is 4.8 in².

The answer is (C).

27. From the problem illustration, the beam clear span, l_n, is the bay width minus the column width.

$$l_n = l - h = 20 \text{ ft} - 1 \text{ ft} = 19 \text{ ft}$$

From ACI 318 Sec. 6.5.4, the shear in end members at the face of first interior supports can be calculated from the equation

$$1.15 w_u\left(\frac{l_n}{2}\right)$$

From ACI 318 Sec. 9.4.3.2, the critical section for shear is at a distance, d, away from the support. Therefore, the shear demand at the critical section is

$$V_u = 1.15 w_u\left(\frac{l_n}{2} - d\right)$$

$$= (1.15)\left(5.4\,\frac{\text{kips}}{\text{ft}}\right)\left(\frac{19 \text{ ft}}{2} - \frac{18 \text{ in}}{12\,\frac{\text{in}}{\text{ft}}}\right)$$

$$= 49.7 \text{ kips}$$

No information about flexural reinforcement was provided, so use ACI 318 Eq. 22.5.5.1 to calculate the shear strength provided by the concrete, V_c. From ACI 318 Table 21.2.1, ϕ is 0.75 for shear. From ACI 318 Table 19.2.4.2, λ is 1.0 for normal weight concrete.

$$\phi V_c = \phi 2\lambda\sqrt{f_c'}\,b_w d$$

$$= \frac{(0.75)(2)(1.0)\sqrt{4000\,\frac{\text{lbf}}{\text{in}^2}}\,(12 \text{ in})(18 \text{ in})}{1000\,\frac{\text{lbf}}{\text{kip}}}$$

$$= 20.5 \text{ kips}$$

This is less than the shear demand, V_u, so shear stirrups are required.

From ACI 318 Sec. 22.5.10.1, the required nominal shear strength provided by the reinforcement, V_s, is

$$V_s = \frac{V_u - \phi V_c}{\phi} = \frac{49.7 \text{ kips} - 20.5 \text{ kips}}{0.75} = 38.9 \text{ kips}$$

From ACI 318 Sec. 22.5.1.2, V_s cannot be taken as greater than

$$V_{s,\max} = 8\sqrt{f_c'}\,b_w d = \frac{(8)\sqrt{4000\,\frac{\text{lbf}}{\text{in}^2}}\,(12 \text{ in})(18 \text{ in})}{1000\,\frac{\text{lbf}}{\text{kip}}}$$

$$= 109 \text{ kips} > 38.9 \text{ kips} \quad [\text{does not control}]$$

The nominal area of no. 3 bars is given in ACI 318 App. A as 0.11 in². From ACI 318 Eq 22.5.10.5.3, the required shear reinforcement spacing, s, is

$$s = \frac{A_v f_{yt} d}{V_s} = \frac{(2)(0.11 \text{ in}^2)\left(60\,\frac{\text{kips}}{\text{in}^2}\right)(18 \text{ in})}{38.9 \text{ kips}}$$

$$= 6.1 \text{ in} \quad (6 \text{ in})$$

Check the minimum spacing, $s_{\min}$, per ACI 318 Sec. 9.7.6.2.2. For $V_s < 4\sqrt{f_c'}\,b_w d$,

$$s_{\min} = \frac{d}{2} = \frac{18 \text{ in}}{2} = 9 \text{ in} < 24 \text{ in} \quad [\text{does not control}]$$

Check the minimum shear reinforcement area, $A_{v,\min}$, per ACI 318 Sec. 9.6.3.3.

$$A_{v,\min} = 0.75\sqrt{f_c'}\left(\frac{b_w s}{f_{yt}}\right)$$

$$= (0.75)\sqrt{4000\,\frac{\text{lbf}}{\text{in}^2}}\left(\frac{(12 \text{ in})(6 \text{ in})}{\left(60\,\frac{\text{kips}}{\text{in}^2}\right)\left(1000\,\frac{\text{lbf}}{\text{kip}}\right)}\right)$$

$$= 0.057 \text{ in}^2 < 0.22 \text{ in}^2 \quad [\text{does not control}]$$

But, per ACI 318 Table 9.6.3.3 the minimum shear reinforcement cannot be less than

$$A_{v,\min} = \frac{50 b_w s}{f_{yt}}$$

$$= \frac{(50)(12 \text{ in})(6 \text{ in})}{\left(60 \frac{\text{kips}}{\text{in}^2}\right)\left(1000 \frac{\text{lbf}}{\text{kip}}\right)}$$

$$= 0.06 \text{ in}^2 < 0.22 \text{ in}^2 \quad [\text{does not control}]$$

(The second equation of ACI 318 Table 9.6.3.3 $A_{v,\min} = 50 b_w s / f_{yt}$, will govern over $A_{v,\min} = (0.75\sqrt{f_c'})(b_w s / f_{yt})$ for $f_{yt} = 60 \text{ kips/in}^2$ and $f_c' < 4444 \text{ lbf/in}^2$.)

The answer is (B).

28. AASHTO Sec. 5.8.4 provides requirements for interface shear reinforcement. Combining AASHTO Eq. 5.8.4.1-1 and Eq. 5.8.4.1-2 and using $\phi = 0.9$ for shear in normal weight concrete from Sec. 5.5.4.2.1 gives

$$V_{ni,\min} = \frac{V_{ui}}{\phi} = \frac{100 \frac{\text{kips}}{\text{ft}}}{0.9} = 111 \text{ kips/ft}$$

Find the area of the concrete, A_{cv}, that is engaged in interface shear transfer. From the problem illustration, $b_{vi} = 14$ in. Because the shear reinforcement per foot length of girder is required, $L_{vi} = 12 \text{ in/ft}$. From AASHTO Eq. 5.8.4.1-6,

$$A_{cv} = b_{vi} L_{vi} = (14 \text{ in})\left(12 \frac{\text{in}}{\text{ft}}\right) = 168 \text{ in}^2/\text{ft}$$

From AASHTO Sec. 5.8.4.3, the cohesion and friction factors for a normal weight, cast-in-place concrete slab on a clean, roughened concrete girder surface are

$$c = 0.28 \text{ kips/in}^2$$
$$\mu = 1.0$$
$$K_1 = 0.3$$
$$K_2 = 1.8 \text{ kips/in}^2$$

Check that $V_{ni,\min}$ does not exceed the limits of AASHTO Eq. 5.8.4.1-4 and Eq. 5.8.4.1-5.

$$K_1 f_c' A_{cv} = (0.3)\left(4 \frac{\text{kips}}{\text{in}^2}\right)\left(168 \frac{\text{in}^2}{\text{ft}}\right)$$

$$= 202 \text{ kips/ft} > V_{ni,\min} \quad [\text{OK}]$$

$$K_2 A_{cv} = \left(1.8 \frac{\text{kips}}{\text{in}^2}\right)\left(168 \frac{\text{in}^2}{\text{ft}}\right)$$

$$= 302 \text{ kips/ft} > V_{ni,\min} \quad [\text{OK}]$$

Therefore, AASHTO Eq. 5.8.4.1-3 controls. Rearranging to solve for the required area of interface shear reinforcement, A_{vf}, gives

$$A_{vf} = \frac{V_{ni,\min} - c A_{cv} - \mu P_c}{\mu f_y}$$

$$= \frac{111 \frac{\text{kips}}{\text{ft}} - \left(0.28 \frac{\text{kips}}{\text{in}^2}\right)\left(168 \frac{\text{in}^2}{\text{ft}}\right) - (1.0)\left(25 \frac{\text{kips}}{\text{ft}}\right)}{(1.0)\left(60 \frac{\text{kips}}{\text{in}^2}\right)}$$

$$= 0.65 \text{ in}^2/\text{ft}$$

The answer is (B).

29. The moment, M, due to the self-weight of the beam is

$$M = \frac{\gamma_c A l^2}{8}$$

$$= \frac{\left(0.145 \frac{\text{kips}}{\text{ft}^3}\right)\left(\dfrac{843 \text{ in}^2}{\left(12 \frac{\text{in}}{\text{ft}}\right)^2}\right)(65 \text{ ft})^2}{8}$$

$$= 448 \text{ ft-kips}$$

The eccentricity, e, of the prestressing force is the distance between the centroid of the section and the centroid of the prestressing strands. From the problem illustration, the distance from the centroid of the prestressing strands to the bottom of the section is

$$2 \text{ in} + \frac{2 \text{ in}}{2} = 3 \text{ in}$$

Therefore, the eccentricity of the prestressing force is

$$e = y_b - 3 \text{ in} = 20.78 \text{ in} - 3 \text{ in} = 17.78 \text{ in}$$

The stress, f_t, at the top of the beam is

$$f_t = \frac{P_e}{A} - \frac{P_e e y_t}{I} + \frac{M y_t}{I}$$

$$= \frac{550 \text{ kips}}{843 \text{ in}^2} - \frac{(550 \text{ kips})(17.78 \text{ in})(21.22 \text{ in})}{203,100 \text{ in}^4}$$

$$+ \frac{(448 \text{ ft-kips})\left(12 \dfrac{\text{in}}{\text{ft}}\right)(21.22 \text{ in})}{203,100 \text{ in}^4}$$

$$= 0.192 \text{ kips/in}^2 \quad (0.19 \text{ kips/in}^2 \text{ (compression)})$$

The answer is (C).

30. For Douglas fir glued laminated timbers (glulam) with the combination symbol 16F-E3, NDS Supp. Table 5A gives a reference design modulus of elasticity, E_x, of

$$E_x = 1,600,000 \text{ lbf/in}^2$$

The moment of inertia, I_{xx}, is

$$I_{xx} = \frac{bd^3}{12} = \frac{(3.5 \text{ in})(7.5 \text{ in})^3}{12} = 123 \text{ in}^4$$

From NDS Sec. 3.5.2, deflection under long-term loading, Δ_T, is given by

$$\Delta_T = K_{cr}\Delta_{LT} + \Delta_{ST}$$

For long-term loads, NDS Sec. 3.5.2 specifies a creep factor, K_{cr}, of 1.5 for structural glued laminated timber used in dry service conditions ($C_M = 1.0$).

The immediate deflection due to the long-term component of the design load, Δ_{LT}, is

$$\Delta_{LT} = \frac{5 w_{LT} l^4}{384 E_x I} = \frac{(5)\left(\dfrac{200 \dfrac{\text{lbf}}{\text{ft}}}{12 \dfrac{\text{in}}{\text{ft}}}\right)\left((10 \text{ ft})\left(12 \dfrac{\text{in}}{\text{ft}}\right)\right)^4}{(384)\left(1,600,000 \dfrac{\text{lbf}}{\text{in}^2}\right)(123 \text{ in}^4)}$$

$$= 0.229 \text{ in}$$

The deflection due to the short-term component of the design load, Δ_{ST}, is

$$\Delta_{ST} = \Delta_{LT}\left(\frac{w_{ST}}{w_{LT}}\right) = (0.229 \text{ in})\left(\frac{100 \dfrac{\text{lbf}}{\text{ft}}}{200 \dfrac{\text{lbf}}{\text{ft}}}\right) = 0.115 \text{ in}$$

The total beam deflection under long-term loading is

$$\Delta_T = K_{cr}\Delta_{LT} + \Delta_{ST} = (1.5)(0.229 \text{ in}) + 0.115 \text{ in}$$
$$= 0.46 \text{ in}$$

The answer is (C).

31. The procedure for calculating the column stability factor, C_P, is covered in NDS Sec. 3.7.1.

The relevant design values for a Douglas fir-south select structural member are

$$F_c = 1050 \text{ lbf/in}^2$$
$$E_{\min} = 440,000 \text{ lbf/in}^2$$

All applicable adjustment factors per NDS Table 4.3.1 are defined as 1.0, except C_P. Therefore, F_c^*, the reference compression design value multiplied by all applicable adjustment factors except C_P, is

$$F_c^* = F_c = 1050 \text{ lbf/in}^2$$

The adjusted modulus of elasticity, $E'_{\min}$, is

$$E'_{\min} = E_{\min} = 440,000 \text{ lbf/in}^2$$

Since it is assumed that $K_e = 1.0$, the effective length of the column, l_e, is equal to the distance between points of lateral support, l. From NDS Sec. 3.7.1.3, the controlling slenderness ratio, l_e/d, is the larger of the ratios l_{e1}/d_1 or l_{e2}/d_2, which are shown in NDS Fig. 3F.

$$\frac{l_e}{d} = \max\begin{cases} \dfrac{l_{e1}}{d_1} = \dfrac{(16 \text{ ft})\left(12 \dfrac{\text{in}}{\text{ft}}\right)}{7.5 \text{ in}} = 25.6 \quad [\text{controls}] \\[4mm] \dfrac{l_{e2}}{d_2} = \dfrac{(8 \text{ ft})\left(12 \dfrac{\text{in}}{\text{ft}}\right)}{5.5 \text{ in}} = 17.5 \end{cases}$$

From NDS Sec. 3.7.1.5, the critical buckling design value for wood compression members, F_{cE}, is

$$F_{cE} = \frac{0.822 E'_{\min}}{\left(\dfrac{l_e}{d}\right)^2} = \frac{(0.822)\left(440,000 \dfrac{\text{lbf}}{\text{in}^2}\right)}{(25.6)^2}$$

$$= 552 \text{ lbf/in}^2$$

The ratio of F_{cE} to F_c^* is

$$\frac{F_{cE}}{F_c^*} = \frac{552 \ \dfrac{\text{lbf}}{\text{in}^2}}{1050 \ \dfrac{\text{lbf}}{\text{in}^2}} = 0.526$$

From NDS Sec. 3.7.1.5, for sawn lumber, $c = 0.8$.

From NDS Eq. 3.7-1, the column stability factor is

$$C_P = \frac{1 + \dfrac{F_{cE}}{F_c^*}}{2c} - \sqrt{\left(\frac{1 + \dfrac{F_{cE}}{F_c^*}}{2c}\right)^2 - \frac{\dfrac{F_{cE}}{F_c^*}}{c}}$$

$$= \frac{1 + 0.526}{(2)(0.8)} - \sqrt{\left(\frac{1 + 0.526}{(2)(0.8)}\right)^2 - \frac{0.526}{0.8}}$$

$$= 0.45$$

The answer is (B).

32. From NDS App. L, Table L2, the critical dimension for the withdrawal of a hex lag screw is the thread length, T, minus the length of the tapered tip, E. For a standard $\frac{1}{2}$ in diameter, 2 in long lag screw, the length of thread penetration not including the tapered tip is

$$p_t = T - E = 1.1875 \text{ in}$$

From NDS Table 11.3.3A, the specific gravity, SG, for Douglas fir-larch (north) is 0.49. (Note that NDS uses G for specific gravity whereas this solution uses SG.)

From NDS Table 11.2A, for a $\frac{1}{2}$ in diameter lag screw with SG $= 0.49$, the reference withdrawal design value per inch of penetration, W, is 367 lbf/in.

The nominal withdrawal design value, Z, is

$$Z = W p_t = \left(367 \ \frac{\text{lbf}}{\text{in}}\right)(1.1875 \text{ in})$$

$$= 436 \text{ lbf} \quad (440 \text{ lbf})$$

The answer is (B).

33. The compressive strength of masonry, f_m', is determined from TMS 602, Sec. 1.4 B.2 Table 2. For type S mortar and CMU with a net area compressive strength of 3250 lbf/in², $f_m' = 2500$ lbf/in².

From TMS 402 Sec. 8.1.6.7.1.1, the minimum lap length for bars in tension is determined by Eq. 8-12, though the length cannot be less than 12 in. There is no transverse reinforcement, so TMS 402 Sec. 8.1.6.7.1.2 does not apply.

TMS 402 Eq. 8-12 gives the required length, l_d.

$$l_d = \frac{0.13 d_b^2 f_y \gamma}{K \sqrt{f_m'}}$$

From TMS 402 Sec. 8.1.6.3, $\gamma = 1.3$ for no. 7 bars.

From ACI 318 App. A, the nominal diameter, d_b, of no. 7 bars is 0.875 in.

From TMS 402 Sec. 8.1.6.3, K is the lesser of the minimum masonry clear cover, the clear spacing between adjacent reinforcement splices, or $9 d_b$.

$$K = \min \begin{cases} \dfrac{h}{2} - \dfrac{d_b}{2} = \dfrac{7.625 \text{ in}}{2} - \dfrac{0.875 \text{ in}}{2} \\ \qquad = 3.375 \text{ in} \quad [\text{controls}] \\ s - 2 d_b = 8 \text{ in} - (2)(0.875 \text{ in}) = 6.25 \text{ in} \\ 9 d_b = (9)(0.875 \text{ in}) = 7.875 \text{ in} \end{cases}$$

The required length is

$$l_d = \frac{0.13 d_b^2 f_y \gamma}{K \sqrt{f_m'}}$$

$$= \frac{(0.13)(0.875 \text{ in})^2 \left(60 \ \dfrac{\text{kips}}{\text{in}^2}\right)\left(1000 \ \dfrac{\text{lbf}}{\text{kip}}\right)(1.3)}{(3.375 \text{ in})\sqrt{2500 \ \dfrac{\text{lbf}}{\text{in}^2}}}$$

$$= 46 \text{ in} > 12 \text{ in} \quad [\text{OK}]$$

The answer is (D).

34. From TMS 402 Sec. 5.2.1.1.1, the span length, l_e, of beams not built integrally with supports is the clear span plus the beam depth, but it need not exceed the distance between the centers of supports. The clear span plus the beam depth is greater than 13 ft, so the distance between the centers of supports controls, and $l_e = 13$ ft.

The moment due to the self-weight of the CMU beam, M_{self}, is

$$M_{\text{self}} = \frac{w_m h l_e^2}{8} = \frac{\left(80 \ \dfrac{\text{lbf}}{\text{ft}^2}\right)(15.625 \text{ in})(13 \text{ ft})^2}{8}$$

$$= 26,400 \text{ in-lbf}$$

The weight, W, of the triangular portion of arching clay masonry that is supported by the lintel is

$$W = w_m \left(\frac{l_e}{2}\right)^2 = \left(80 \ \frac{\text{lbf}}{\text{ft}^2}\right)\left(\frac{13 \text{ ft}}{2}\right)^2 = 3380 \text{ lbf}$$

The maximum moment due to the triangular loading, M_{arch}, can be found from *AISC Manual* Table 3-23, case 3.

$$M_{arch} = \frac{Wl_e}{6} = \frac{(3380 \text{ lbf})(13 \text{ ft})\left(12 \ \dfrac{\text{in}}{\text{ft}}\right)}{6}$$
$$= 87,880 \text{ in-lbf}$$

The total moment, M_{tot}, is

$$M_{tot} = M_{self} + M_{arch} = 26,400 \text{ in-lbf} + 87,800 \text{ in-lbf}$$
$$= 114,200 \text{ in-lbf}$$

The procedure for determining the stresses in a reinforced masonry section is outlined in many structural engineering textbooks.

From TMS 402 Sec. 4.2.2.1, the modulus of elasticity of steel reinforcement, E_s, is

$$E_s = 29,000,000 \text{ lbf/in}^2$$

From TMS 402 Sec. 4.2.2.2.1, the modulus of elasticity for concrete masonry, E_m, is

$$E_m = 900f'_m = (900)\left(2000 \ \frac{\text{lbf}}{\text{in}^2}\right) = 1,800,000 \text{ lbf/in}^2$$

The modular ratio, n, is

$$n = \frac{E_s}{E_m} = \frac{29,000,000 \ \dfrac{\text{lbf}}{\text{in}^2}}{1,800,000 \ \dfrac{\text{lbf}}{\text{in}^2}} = 16.11$$

From ACI 318 App. A, the nominal area of a no. 5 bar is 0.31 in². The tension reinforcement ratio, ρ, for two bars is

$$\rho = \frac{A_s}{b_w d} = \frac{(2)(0.31 \text{ in}^2)}{(7.625 \text{ in})(13 \text{ in})} = 0.006255$$
$$\rho n = (0.006255)(16.11) = 0.1008$$

The neutral axis depth factor, k, is

$$k = \sqrt{2\rho n + (\rho n)^2} - \rho n$$
$$= \sqrt{(2)(0.1008) + (0.1008)^2} - 0.1008$$
$$= 0.359$$

The lever-arm factor, j, is

$$j = 1 - \frac{k}{3} = 1 - \frac{0.359}{3} = 0.880$$

The maximum compression stress, f_b, is

$$f_b = \frac{2M_{tot}}{jkb_w d^2} = \frac{(2)(114,200 \text{ in-lbf})}{(0.880)(0.359)(7.625 \text{ in})(13 \text{ in})^2}$$
$$= 561 \text{ lbf/in}^2 \quad (560 \text{ lbf/in}^2)$$

The answer is (B).

35. From TMS 402 Sec. 5.3.1.4(b), the vertical spacing cannot exceed 16 longitudinal bar diameters, 48 lateral tie bar or wire diameters, or the least cross-sectional dimension of the member. The nominal diameters of no. 3 and no. 6 bars can be found in ACI 318 App. A. The minimum required vertical spacing of the lateral ties, s_{min}, is

$$s_{min} = \min \begin{cases} (16)(0.75 \text{ in}) = 12 \text{ in} & \text{[controls]} \\ (48)(0.375 \text{ in}) = 18 \text{ in} \\ 15.625 \text{ in} \end{cases}$$

The answer is (A).

36. The end bearing capacity, P_{end}, is

$$P_{end} = q_{end}\left(\frac{\pi d^2}{4}\right) = \left(\frac{10,000 \ \dfrac{\text{lbf}}{\text{ft}^2}}{1000 \ \dfrac{\text{lbf}}{\text{kip}}}\right)\left(\frac{\pi\left(\dfrac{18 \text{ in}}{12 \ \dfrac{\text{in}}{\text{ft}}}\right)^2}{4}\right)$$
$$= 17.7 \text{ kips}$$

The skin-friction capacity, P_{skin}, is

$$P_{skin} = q_{skin}\pi dL_{firm}$$
$$= \left(\frac{1000 \ \dfrac{\text{lbf}}{\text{ft}^2}}{1000 \ \dfrac{\text{lbf}}{\text{kip}}}\right)\pi\left(\frac{18 \text{ in}}{12 \ \dfrac{\text{in}}{\text{ft}}}\right)(15 \text{ ft})$$
$$= 70.7 \text{ kips}$$

The drag-down force, P_{drag}, from the soft soil is

$$P_{drag} = q_{drag}\pi dL_{soft}$$
$$= \left(\frac{400 \ \dfrac{\text{lbf}}{\text{ft}^2}}{1000 \ \dfrac{\text{lbf}}{\text{kip}}}\right)\pi\left(\frac{18 \text{ in}}{12 \ \dfrac{\text{in}}{\text{ft}}}\right)(10 \text{ ft})$$
$$= 18.8 \text{ kips}$$

The total allowable axial pile load, P, is

$$
\begin{aligned}
P &= P_{\text{end}} + P_{\text{skin}} - P_{\text{drag}} \\
&= 17.7 \text{ kips} + 70.7 \text{ kips} - 18.8 \text{ kips} \\
&= 69.6 \text{ kips} \quad (70 \text{ kips})
\end{aligned}
$$

The answer is (C).

37. The factored column load, P_u, is

$$
\begin{aligned}
P_u &= 1.2D + 1.6L \\
&= (1.2)(100 \text{ kips}) + (1.6)(60 \text{ kips}) \\
&= 216 \text{ kips}
\end{aligned}
$$

The factored footing self-weight, $W_{u,\text{footing}}$, is

$$
\begin{aligned}
W_{u,\text{footing}} &= 1.2\gamma_c B^2 t_{\text{footing}} \\
&= \dfrac{(1.2)\left(150 \ \dfrac{\text{lbf}}{\text{ft}^3}\right)(7 \text{ ft})^2 (3 \text{ ft})}{1000 \ \dfrac{\text{lbf}}{\text{kip}}} \\
&= 26.5 \text{ kips}
\end{aligned}
$$

The factored strap beam self-weight, $W_{u,\text{strap}}$, is

$$
\begin{aligned}
W_{u,\text{strap}} &= 1.2\gamma_c lwt_{\text{strap}} \\
&= \dfrac{(1.2)\left(150 \ \dfrac{\text{lbf}}{\text{ft}^3}\right)(20 \text{ ft})(2 \text{ ft})(2.5 \text{ ft})}{1000 \ \dfrac{\text{lbf}}{\text{kip}}} \\
&= 18 \text{ kips}
\end{aligned}
$$

Due to symmetry, the reactions under each footing are the same.

$$
\begin{aligned}
R_u &= P_u + W_{u,\text{footing}} + \dfrac{W_{u,\text{strap}}}{2} \\
&= 216 \text{ kips} + 26.5 \text{ kips} + \dfrac{18 \text{ kips}}{2} \\
&= 251.5 \text{ kips}
\end{aligned}
$$

Take moments about the beam/footing interface. The moment, M_u, at the strap/footing interface is

$$
\begin{aligned}
M_u &= P_u(B - a) - (R_u - W_{u,\text{footing}})\left(\dfrac{B}{2}\right) \\
&= (216 \text{ kips})(7 \text{ ft} - 1 \text{ ft}) \\
&\quad - (251.5 \text{ kips} - 26.5 \text{ kips})\left(\dfrac{7 \text{ ft}}{2}\right) \\
&= 510 \text{ ft-kips}
\end{aligned}
$$

The answer is (B).

38. The at-rest pressure coefficient, K_o, is used in the calculation of soil and surcharge loads for basement walls restrained from movement.

For retaining wall questions, it can be helpful to draw the forces acting on the wall.

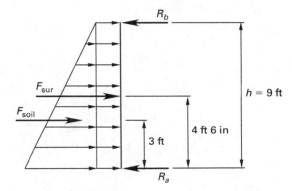

The resultant from the surcharge load, F_{sur}, is

$$
F_{\text{sur}} = qK_o h = \left(100 \ \dfrac{\text{lbf}}{\text{ft}^2}\right)(0.50)(9 \text{ ft}) = 450 \text{ lbf/ft}
$$

The resultant from the soil load, F_{soil}, is

$$
\begin{aligned}
F_{\text{soil}} &= \tfrac{1}{2}\gamma_s K_o h^2 = \left(\dfrac{1}{2}\right)\left(110 \ \dfrac{\text{lbf}}{\text{ft}^3}\right)(0.50)(9 \text{ ft})^2 \\
&= 2228 \text{ lbf/ft}
\end{aligned}
$$

The horizontal reaction, R_a, at the basement slab on grade is

$$
\begin{aligned}
R_a &= \tfrac{1}{2}F_{\text{sur}} + \tfrac{2}{3}F_{\text{soil}} \\
&= \left(\dfrac{1}{2}\right)\left(450 \ \dfrac{\text{lbf}}{\text{ft}}\right) + \left(\dfrac{2}{3}\right)\left(2228 \ \dfrac{\text{lbf}}{\text{ft}}\right) \\
&= 1700 \text{ lbf/ft}
\end{aligned}
$$

The answer is (D).

39. ACI 318 Sec. 26.12.4 describes permissible actions for investigating low-strength test results.

Per ACI 318 Sec. 26.12.4.1(b), if analysis can establish that the load-carrying capacity of the structure is not jeopardized even when considering a concrete strength lower than the specified design strength, further investigation is not required.

Per ACI 318 Sec. 26.12.4.1(c), if low-strength concrete is confirmed, tests of cores drilled in the questionable area may be permitted in accordance with ASTM C42.

Per ACI 318 Sec. 26.12.4.1(f), if doubt remains about the structural adequacy of the slab, strength evaluations may be conducted in accordance with ACI 318 Chap. 27.

The answer is (D).

40. Per IBC Sec. 1705.6 and Table 1705.6, continuous inspections are required to verify the use of proper materials, densities, and lift thicknesses during placement and compaction of compacted fill. The other listed tasks require only periodic inspections.

The answer is (C).

Solutions
Vertical Forces Component: Buildings Depth Module Exam

41. (a) The design of two-way slabs using the direct design method is covered in ACI 318 Sec. 8.10. The slab satisfies all of the listed limitations of ACI 318 Sec. 8.10.2, so the direct design method may be used. Per the problem statement, the direct design method should be used; therefore, it can be assumed that the limitations are satisfied and it is not necessary to check for them.

From the problem illustration, the span length parallel to the direction that the moments are being determined is

$$l_1 = 16 \text{ ft}$$

The clear span measured face-to-face of supports is the span length minus the column dimension.

$$l_n = 16 \text{ ft} - 1 \text{ ft} = 15 \text{ ft}$$

From the problem illustration, the span length perpendicular to the direction that the moments are being determined is

$$l_2 = 20 \text{ ft}$$

The given loads must be factored for concrete strength design per ACI 318 Sec. 5.3 or IBC Sec. 1605.2. By inspection, the load combination from ACI 318 Eq. 5.3.1b or IBC Eq. 16-2 $(1.2D + 1.6L)$ will control for the design of the slab. The factored area load, q_u, is

$$q_u = 1.2q_{\text{dead}} + 1.6q_{\text{live}}$$
$$= \frac{(1.2)\left(140 \ \dfrac{\text{lbf}}{\text{ft}^2}\right) + (1.6)\left(100 \ \dfrac{\text{lbf}}{\text{ft}^2}\right)}{1000 \ \dfrac{\text{lbf}}{\text{kip}}}$$
$$= 0.328 \text{ kips/ft}^2$$

From ACI 318 Eq. 8.10.3.2, the total factored moment, M_o, per span is

$$M_o = \frac{q_u l_2 l_n^2}{8} = \frac{\left(0.328 \ \dfrac{\text{kips}}{\text{ft}^2}\right)(20 \text{ ft})(15 \text{ ft})^2}{8}$$
$$= 184.5 \text{ ft-kips} \quad (185 \text{ ft-kips})$$

Use the clear span dimension of 15 ft, not the column centerline spacing of 16 ft, in the calculation of the total factored moment.

From ACI 318 Sec. 8.10.4.2, distribute the total factored moment in the end spans.

The exterior negative moment for the end spans, $M_{ue,e}^-$ is

$$M_{ue,e}^- = 0.26M_o = (0.26)(184.5 \text{ ft-kips})$$
$$= 48 \text{ ft-kips}$$

The positive moment for the end spans, $M_{u,e}^+$, is

$$M_{u,e}^+ = 0.52M_o = (0.52)(184.5 \text{ ft-kips})$$
$$= 96 \text{ ft-kips}$$

The interior negative moment for the end spans, $M_{ui,e}^-$, is

$$M_{ui,e}^- = 0.70M_o = (0.70)(184.5 \text{ ft-kips})$$
$$= 129 \text{ ft-kips}$$

From ACI 318 Sec. 8.10.4.1, distribute the total factored moment in the interior spans.

The positive moment for interior spans, $M_{u,i}^+$, is

$$M_{u,i}^+ = 0.35M_o = (0.35)(184.5 \text{ ft-kips})$$
$$= 65 \text{ ft-kips}$$

The negative moment for interior spans, $M_{u,i}^-$, is

$$M_{u,i}^- = 0.65M_o = (0.65)(184.5 \text{ ft-kips})$$
$$= 120 \text{ ft-kips}$$

From ACI 318 Sec. 8.10.5, distribute a portion of the moments to the column strip. The values α_{f1} and β_t are ratios of beam to slab stiffness. For a flat plate slab with no beams, the beam stiffness is zero. Therefore, α_{f1} and β_t are 0.

$$\frac{l_2}{l_1} = \frac{20 \text{ ft}}{16 \text{ ft}} = 1.25$$

And,

$$\frac{\alpha_{f1} l_2}{l_1} = \frac{(0)(20 \text{ ft})}{16 \text{ ft}} = 0$$

The tables in ACI 318 Sec. 8.10.5.1, Sec. 8.10.5.2, and Sec. 13.6.4.4 give percentages of factored moments that are resisted by the column strip.

From Sec. 8.10.5.2, for column strip end spans, the exterior negative moment is 100% of $M_{ue,e}^-$.

$$M_{ue,ce}^- = 1 M_{ue,e}^- = (1)(48 \text{ ft-kips}) = \boxed{48 \text{ ft-kips}}$$

From Sec. 8.10.5.5, for column strip end spans, the positive moment is 60% of $M_{u,e}^+$.

$$M_{u,ce}^+ = 0.60 M_{u,e}^+ = (0.60)(96 \text{ ft-kips}) = \boxed{58 \text{ ft-kips}}$$

From Sec. 8.10.5.1, for column strip end spans, the interior negative moment is 75% of $M_{ui,e}^-$.

$$M_{ui,ce}^- = 0.75 M_{ui,e}^- = (0.75)(129 \text{ ft-kips}) = \boxed{97 \text{ ft-kips}}$$

From Sec. 8.10.5.5, for column strip interior spans, the positive moment is 60% of $M_{u,i}^+$.

$$M_{u,ci}^+ = 0.60 M_{u,i}^+ = (0.60)(65 \text{ ft-kips}) = \boxed{39 \text{ ft-kips}}$$

From Sec. 8.10.5.1, for column strip interior spans, the negative moment is 75% of $M_{u,i}^-$.

$$M_{u,ci}^- = 0.75 M_{u,i}^- = (0.75)(120 \text{ ft-kips}) = \boxed{90 \text{ ft-kips}}$$

(Note that the presented solution ignores the modification of factored moments per ACI 318 Sec. 8.10.4.3. It is possible to modify the negative and positive factored moments under the provision of this section, though this would require more steps than is necessary to solve this problem.)

41. (b) The moment diagram for the column strip in part 41(a) is shown.

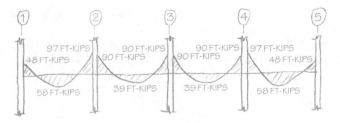

41. (c) From ACI 318 Sec. 8.4.1.5, column strip width is the smaller of $0.25 l_1$ or $0.25 l_2$ on each side of the column centerline. l_1 is the controlling dimension, since $l_1 < l_2$.

The column strip width, b, is

$$b = (2)(0.25 l_1) = (2)(0.25)(16 \text{ ft})\left(12 \frac{\text{in}}{\text{ft}}\right) = 96 \text{ in}$$

Determine the required flexural reinforcement. (This solution presents only one of many acceptable solving methods, which are further described in code books and concrete textbooks.)

Use the end span moment values from part 41(a). Because $M_{ue,ce}^- = 48$ ft-kips and $M_{u,ce}^+ = 58$ ft-kips are similar in magnitude, it is reasonable to use the same reinforcement design for both locations.

Therefore, the design factored moment, M_u, for the end span exterior negative moment and the end span positive moment is

$$M_{u,ce}^+ = M_u = 58 \text{ ft-kips}$$

Calculate $M_u/(f_c' b d^2)$ and determine the tension reinforcement index, ω, using a design table.[1]

$$\frac{M_u}{f_c' b d^2} = \frac{(58 \text{ ft-kips})\left(12 \dfrac{\text{in}}{\text{ft}}\right)}{\left(\dfrac{4000 \dfrac{\text{lbf}}{\text{in}^2}}{1000 \dfrac{\text{lbf}}{\text{kip}}}\right)(96 \text{ in})(7.5 \text{ in})^2} = 0.0322$$

Using a design table that calculates $M_u/(f_c' b d^2)$ against ω,

$$\omega = 0.0365$$

The required reinforcement ratio, ρ, is

$$\rho = \frac{\omega f_c'}{f_y} = \frac{(0.0365)\left(4000 \dfrac{\text{lbf}}{\text{in}^2}\right)}{60{,}000 \dfrac{\text{lbf}}{\text{in}^2}} = 0.00243$$

Check to see if the section is tension controlled. For $f_c' = 4000$ lbf/in^2, ACI 318 Sec. 22.2.2.4.3 gives $\beta_1 =$

[1]Acceptable design tables may be found in the *ACI Design Handbook* (SP-17), *Notes on ACI 318: Building Code Requirements for Structural Concrete*, or *Structural Engineering Reference Manual*. Bibliographic information is provided for these references in this book's Codes and References.

0.85. The limiting reinforcement ratio for a tension-controlled section is

$$\rho_t = 0.319\beta_1 \left(\frac{f_c'}{f_y}\right) = (0.319)(0.85) \left(\frac{4000 \, \frac{\text{lbf}}{\text{in}^2}}{60{,}000 \, \frac{\text{lbf}}{\text{in}^2}}\right)$$

$$= 0.0181 > 0.00243 \quad \text{[tension-controlled]}$$

The required reinforcement area, A_s, is

$$A_s = \rho bd = (0.00243)(96 \text{ in})(7.5 \text{ in}) = 1.75 \text{ in}^2$$

From ACI 318 Sec. 8.6.1.1, the minimum area of flexural reinforcement steel, $A_{s,\text{min}}$, is

$$A_{s,\text{min}} = 0.0018 bt_s = (0.0018)(96 \text{ in})(9 \text{ in})$$

$$= 1.56 \text{ in}^2 < 1.75 \text{ in}^2 \quad \text{[does not control]}$$

Try 6 no. 5 bars. From ACI 318 App. A, the nominal area of no. 5 bars is 0.31 in^2.

$$A_{s,\text{provided}} = A_{\text{bar}}(\text{no. of bars}) = (0.31 \text{ in}^2)(6)$$

$$= 1.86 \text{ in}^2 > 1.75 \text{ in}^2 \quad \text{[OK]}$$

From ACI 318 Sec. 8.7.2.2, the minimum bar spacing, s_{min}, is the lesser of $2 \times$ slab thickness and 18 in at critical sections

$$s_{\text{min}} = 2t_s = (2)(9 \text{ in}) = 18 \text{ in}$$

Check the minimum bar spacing.

$$s = \frac{b}{\text{no. of bars}} = \frac{96 \text{ in}}{6} = 16 \text{ in} < 18 \text{ in} \quad \text{[OK]}$$

For the end span exterior negative moment region, provide six no. 5 top bars in the column strip.

For the end span positive moment region, provide six no. 5 top bars in the column strip.

Use the value for the column strip end span interior negative moment, $M_{ui,ce}^-$, from part 41(a) for the first interior column.

$$M_{ui,ce}^- = M_u = 97 \text{ ft-kips}$$

Calculate $M_u/(f_c'bd^2)$, and determine ω using a design table.

$$\frac{M_u}{f_c'bd^2} = \frac{(97 \text{ ft-kips})\left(12 \, \frac{\text{in}}{\text{ft}}\right)}{\left(\dfrac{4000 \, \frac{\text{lbf}}{\text{in}^2}}{1000 \, \frac{\text{lbf}}{\text{kip}}}\right)(96 \text{ in})(7.5 \text{ in})^2} = 0.0539$$

Using a design table that calculates $M_u/(f_c'bd^2)$ against ω,

$$\omega = 0.0621$$

The required reinforcement ratio, ρ, is

$$\rho = \frac{\omega f_c'}{f_y} = \frac{(0.0621)\left(4000 \, \frac{\text{lbf}}{\text{in}^2}\right)}{60{,}000 \, \frac{\text{lbf}}{\text{in}^2}} = 0.00414$$

Check to see if the section is tension controlled.

$$\rho_t = 0.0181 > 0.00414 \quad \text{[tension-controlled]}$$

The required reinforcement area, A_s, is

$$A_s = \rho bd = (0.00414)(96 \text{ in})(7.5 \text{ in}) = 2.98 \text{ in}^2$$

By inspection, minimum steel requirements of ACI 318 Sec. 8.6.1.1 do not control.

Try 10 no. 5 bars.

$$A_{s,\text{provided}} = A_{\text{bar}}(\text{no. of bars}) = (0.31 \text{ in}^2)(10)$$

$$= 3.10 \text{ in}^2 > 2.98 \text{ in}^2 \quad \text{[OK]}$$

Check the minimum bar spacing.

$$s = \frac{b}{\text{no. of bars}} = \frac{96 \text{ in}}{10} = 9.6 \text{ in} < 18 \text{ in} \quad \text{[OK]}$$

At the first interior column, provide 10 no. 5 top bars in the column strip.

(Note that this solution presents only one acceptable configuration for flexural reinforcement. Other bar sizes and spacing would also be acceptable, provided that the moment demand is satisfied and the bar spacing does not exceed the code requirements. Although it does not control, a complete solution should include the checks for minimum steel area.)

An elevation of the column strip end span that uses the minimum bar extents given in ACI 318 Fig. 8.7.4.1.3(a) is shown.

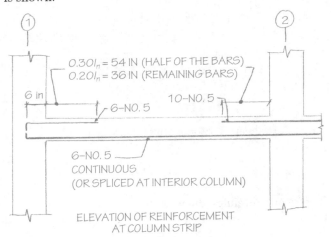

ELEVATION OF REINFORCEMENT
AT COLUMN STRIP

41. (d) For slab-to-column connections, the assumed distribution of shear stress is given in ACI 318 Fig. 8.4.4.2.3. Considering the contribution from the factored shear force, V_u, and the unbalanced factored moment, M_u, the maximum factored shear stress, v_u, is given in ACI 318 Fig. 8.4.4.2.3 as

$$v_{u,\mathrm{AB}} = v_{ug} + \frac{\gamma_v M_u c_{\mathrm{AB}}}{J}$$

Using ACI 318 Fig. R8.4.4.2.3(b) as a reference, the dimensions of the critical section for punching shear can be drawn as shown. (This sketch is not required by the solution and is shown for reference only.)

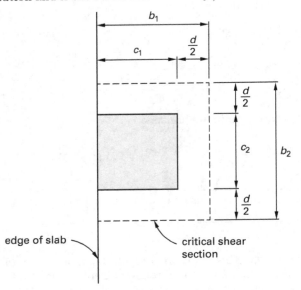

The length of the critical section parallel to the direction of the span, b_1, is

$$b_1 = c_1 + \frac{d}{2} = 12 \text{ in} + \frac{7.5 \text{ in}}{2}$$
$$= 15.8 \text{ in}$$

The length of the critical section perpendicular to the direction of the span, b_2, is

$$b_2 = c_2 + d = 12 \text{ in} + 7.5 \text{ in}$$
$$= 19.5 \text{ in}$$

The perimeter of the critical section for shear, b_o, is

$$b_o = 2b_1 + b_2 = (2)(15.8 \text{ in}) + 19.5 \text{ in}$$
$$= 51 \text{ in}$$

The area of concrete, A_c, resisting shear transfer is

$$A_c = b_o d = (51 \text{ in})(7.5 \text{ in})$$
$$= 382.5 \text{ in}^2$$

The tributary area, A_t, of column B-1 is

$$A_t = l_2\left(\frac{l_1}{2} + \frac{c_1}{2}\right) = (20 \text{ ft})\left(\frac{16 \text{ ft}}{2} + \frac{12 \text{ in}}{(2)\left(12 \frac{\text{in}}{\text{ft}}\right)}\right)$$
$$= 170 \text{ ft}^2$$

Use $q_u = 0.328 \text{ kips/ft}^2$ from part 41(a). Excluding the area inside the critical shear section, the factored shear force, V_u, is

$$V_u = q_u(A_t - b_1 b_2)$$
$$= \left(0.328 \frac{\text{kips}}{\text{ft}^2}\right)\left(170 \text{ ft}^2 - \frac{(15.8 \text{ in})(19.5 \text{ in})}{\left(12 \frac{\text{in}}{\text{ft}}\right)^2}\right)$$
$$= 55.1 \text{ kips}$$

(Note that excluding the area inside the critical shear section in the calculation of the shear demand is more accurate, but is not required.)

The unbalanced moment, M_u, that must be transferred

between the slab and the edge column is given in ACI 318 Sec. 8.10.7.3. M_o was determined in part 41(a).

$$M_u = 0.3M_o$$

$$= (0.3)(184.5 \text{ ft-kips})\left(1000 \frac{\text{lbf}}{\text{kip}}\right)\left(12 \frac{\text{in}}{\text{ft}}\right)$$

$$= 664{,}200 \text{ in-lbf}$$

From ACI 318 Eq. 8.4.2.3.2, the portion of the unbalanced moment transferred by flexure, γ_f, is

$$\gamma_f = \frac{1}{1 + \dfrac{2}{3}\sqrt{\dfrac{b_1}{b_2}}} = \frac{1}{1 + \left(\dfrac{2}{3}\right)\sqrt{\dfrac{15.8 \text{ in}}{19.5 \text{ in}}}} = 0.625$$

From ACI 318 Eq. 8.4.4.2.2, the remainder of the unbalanced moment is transferred by shear.

$$\gamma_v = 1 - \gamma_f = 1 - 0.625 = 0.375$$

From ACI 318 Sec. R8.4.4.2.3 and PCA Fig. 16-13, the polar moment of inertia, J, divided by the distance to the neutral axis for an edge column, c_{AB}, is

$$\frac{J}{c_{AB}} = \frac{2b_1^2 d(b_1 + 2b_2) + d^3(2b_1 + b_2)}{6b_1}$$

$$= \frac{\begin{array}{c}(2)(15.8 \text{ in})^2(7.5 \text{ in})\big(15.8 \text{ in} + (2)(19.5 \text{ in})\big) \\ + (7.5 \text{ in})^3\big((2)(15.8 \text{ in}) + 19.5 \text{ in}\big)\end{array}}{(6)(15.8 \text{ in})}$$

$$= 2392 \text{ in}^3$$

The maximum factored shear stress, v_u, is

$$v_u = \frac{V_u}{A_c} + \frac{\gamma_v M_u c_{AB}}{J}$$

$$= \frac{(55.1 \text{ kips})\left(1000 \dfrac{\text{lbf}}{\text{kip}}\right)}{382.5 \text{ in}^2} + \frac{(0.375)(664{,}200 \text{ in-lbf})}{2383 \text{ in}^3}$$

$$= 248 \text{ lbf/in}^2$$

From ACI 318 Sec. 22.6.5.2, for nonprestressed slabs and footings, the stress corresponding to the nominal shear strength provided by the slab, v_c, is the smallest of the values from ACI 318 Table 22.6.5.2. For square columns ($\beta = 1.0$), case (b) will not control. For edge columns ($\alpha_s = 30$), case (c) will not control for values where $\alpha_s d/b_o > 2$.

$$\frac{\alpha_s d}{b_o} = \frac{(30)(7.5 \text{ in})}{51 \text{ in}} = 4.41 > 2 \quad \text{[does not control]}$$

Therefore, ACI 318 Table 22.6.5.2 case (a) will control, and the stress corresponding to the nominal shear strength provided by concrete, v_c, is

$$v_c = 4\lambda\sqrt{f_c'}$$

From ACI Table 19.2.4.2, λ is 1.0 for normal weight concrete. From ACI 318 Table 21.2.1, ϕ is 0.75 for shear.

From ACI Eq. 22.6.1.2, for slabs without shear reinforcement, the maximum shear stress cannot exceed ϕv_n.

Compare ϕv_n to v_u.

$$\phi v_n = \phi 4\lambda\sqrt{f_c'} = (0.75)(4)(1.0)\sqrt{4000 \frac{\text{lbf}}{\text{in}^2}}$$

$$= \boxed{190 \text{ lbf/in}^2 < 248 \text{ lbf/in}^2 \quad \text{[not OK]}}$$

Therefore,

$$\boxed{\text{the slab is inadequate for punching shear.}}$$

41. (e) Two ways that the design could be modified to address inadequate punching shear strength include

> • providing drop panels
> • providing headed shear stud reinforcement

Other correct modifications include

- providing beams at the column lines

- using higher strength concrete

- using a thicker slab

- providing column capitals

42. ASD Solution

42. (a) The procedure for analyzing composite flexural members is given by AISC 360 Sec. I3.

From AISC 360 Sec. I3-1a, for an interior beam, the effective slab width, b, is the lesser of one-quarter of the span or the beam spacing.

$$b = \min\begin{cases} \dfrac{L}{4} = \left(\dfrac{30 \text{ ft}}{4}\right)\left(12 \dfrac{\text{in}}{\text{ft}}\right) = 90 \text{ in} \quad \text{[controls]} \\[2ex] s = \dfrac{B}{3} = \left(\dfrac{25 \text{ ft}}{3}\right)\left(12 \dfrac{\text{in}}{\text{ft}}\right) = 100 \text{ in} \end{cases}$$

For the case of the metal deck spanning perpendicular to the beam span, only the concrete above the metal deck will be effective. Therefore, the area of concrete slab within the effective slab width, A_c, is

$$A_c = bt_s = (90 \text{ in})(3 \text{ in}) = 270 \text{ in}^2$$

From AISC *Steel Construction Manual* Table 1-1, the area, A_s, of a W16 × 26 beam is

$$A_s = 7.68 \text{ in}^2$$

From AISC 360 Sec. I3-2d(1), the total horizontal shear, V', for a fully composite beam is the lesser of AISC 360 Eq. I3-1a or Eq. I3-1b.

$$V' = \min \begin{cases} 0.85 f'_c A_c = (0.85)\left(4 \dfrac{\text{kips}}{\text{in}^2}\right)(270 \text{ in}^2) \\ \qquad = 918 \text{ kips} \\ F_y A_s = \left(50 \dfrac{\text{kips}}{\text{in}^2}\right)(7.68 \text{ in}^2) \\ \qquad = 384 \text{ kips} \quad \text{[controls]} \end{cases}$$

From AISC *Steel Construction Manual* Table 3-21, for a perpendicular deck and one ³⁄₄ in diameter strong stud per rib with 4 kips/in² normal-weight concrete, the nominal stud shear strength, Q_n, is

$$Q_n = 21.5 \text{ kips/stud}$$

To provide complete shear connection and full composite action, the required number of connectors on either side of the point of maximum moment is $n = V'/Q_n$.

For a simply supported beam, V'/Q_n must be multiplied by two to get the total number of required shear studs along the length of the beam. The required number of shear studs for full composite action is

$$2n = \frac{2V'}{Q_n} = \frac{(2)(384 \text{ kips})}{21.5 \dfrac{\text{kips}}{\text{stud}}} = \boxed{35.7 \text{ studs} \quad (36 \text{ studs})}$$

Check flexural strength.

Considering only the provided dead and live loads, from IBC Eq. 16-9, the distributed beam load, w, is

$$w = w_D + w_L = \frac{750 \dfrac{\text{lbf}}{\text{ft}} + 1000 \dfrac{\text{lbf}}{\text{ft}}}{1000 \dfrac{\text{lbf}}{\text{kip}}}$$

$$= 1.75 \text{ kips/ft}$$

For a simply supported beam, the required flexural strength, M_r, is

$$M_r = \frac{wL^2}{8} = \frac{\left(1.75 \dfrac{\text{kips}}{\text{ft}}\right)(30 \text{ ft})^2}{8}$$

$$= 197 \text{ ft-kips}$$

Following the procedure from AISC *Steel Construction Manual*, Part 3, the depth of the compression block, a, is

$$a = \frac{F_y A_s}{0.85 f'_c b} = \frac{\left(50 \dfrac{\text{kips}}{\text{in}^2}\right)(7.68 \text{ in}^2)}{(0.85)\left(4 \dfrac{\text{kips}}{\text{in}^2}\right)(90 \text{ in})}$$

$$= 1.25 \text{ in}$$

The compression block is within the slab thickness ($a < Y_{\text{con}}$), so the distance from the top flange of the composite steel beam to the plastic neutral axis is $Y1 = 0$ in per AISC *Steel Construction Manual* Table 3-19.

The distance, $Y2$, from the concrete flange force to the steel beam top flange is

$$Y2 = Y_{\text{con}} - \frac{a}{2} = 6 \text{ in} - \frac{1.25 \text{ in}}{2}$$

$$= 5.38 \text{ in}$$

Interpolating from AISC *Steel Construction Manual* Table 3-19, for a W16 × 26 beam,

$$\frac{M_p}{\Omega_b} = \boxed{253 \text{ ft-kips} > 197 \text{ ft-kips} \quad \text{[OK]}}$$

Therefore, the composite beam is adequate in flexure.

Check deflection.

The allowable deflection limit, $\Delta_{\text{allowable}}$, is

$$\Delta_{\text{allowable}} = \frac{L}{360} = \frac{(30 \text{ ft})\left(12 \dfrac{\text{in}}{\text{ft}}\right)}{360}$$

$$= 1 \text{ in}$$

Use the lower bound elastic moment of inertia, I_{LB}, to calculate the live load deflection. Interpolating from

AISC *Steel Construction Manual* Table 3-20, for a W16 × 26 beam, $I_{LB} = 972$ in^4.

$$\Delta_L = \frac{5 w_L L^4}{384 EI}$$

$$= \frac{(5)\left(\dfrac{1000 \ \frac{\text{lbf}}{\text{ft}}}{\left(1000 \ \frac{\text{lbf}}{\text{kip}}\right)\left(12 \ \frac{\text{in}}{\text{ft}}\right)}\right)\left((30 \text{ ft})\left(12 \ \frac{\text{in}}{\text{ft}}\right)\right)^4}{(384)\left(29{,}000 \ \dfrac{\text{kips}}{\text{in}^2}\right)(972 \text{ in}^4)}$$

$$= \boxed{0.65 \text{ in} < 1 \text{ in} \quad [\text{OK}]}$$

> The composite beam meets the deflection criteria.

42. (b) The beam reaction, R_a, is

$$R_a = \frac{wL}{2}$$

$$= \frac{\left(1.75 \ \dfrac{\text{kips}}{\text{ft}}\right)(30 \text{ ft})}{2}$$

$$= 26.3 \text{ kips}$$

Available strengths for single-plate connections are given in AISC *Steel Construction Manual* Table 10-10a. Using the table, the most efficient connection design with an available strength, R_n/Ω, greater than the beam reaction, R_a, is a connection with three A325 bolts ($n = 3$, $L = 8.5$ in) and a plate thickness of $t_p = {}^5\!/_{16}$ in. A325 bolts are in bolt group A. Conservatively assume condition of threads included (N) in the faying surface. From the bottom of AISC *Steel Construction Manual* Table 10-10a, the weld size for the plate to girder connection is $\frac{1}{4}$ in.

From AISC *Steel Construction Manual* Table 10-10a, the available connection strength is

$$\frac{R_n}{\Omega} = 28.8 \text{ kips} > 26.3 \text{ kips} \quad [\text{OK}]$$

From the AISC *Steel Construction Manual* description of Table 10-10a, the tabulated strengths consider the limit-states of bolt shear, bolt bearing on the plate, shear yielding of the plate, shear rupture of the plate, block shear rupture of the plate, and weld shear. However, AISC *Steel Construction Manual* Table 10-10a does not include a check for bolt bearing on the beam web. From AISC *Steel Construction Manual* Table 1-1, the beam web thickness, t_w, is 0.25 in, which is less than the plate thickness. Therefore, bolt bearing on the beam web should be checked per AISC 360 Sec. J3-10.

Considering deformations at the bolt holes under service loads, use AISC 360 Eq. J3-6a.

The clear distance, L_c, between the edge of the hole and the edge of an adjacent hole in the beam web is the bolt spacing minus the bolt hole diameter. From AISC 360 Table J3.3, the diameter of a standard hole for a $\frac{3}{4}$ in diameter bolt is $^{13}\!/_{16}$ in. Therefore,

$$L_c = 3 \text{ in} - \frac{13}{16} \text{ in}$$

$$= 2.19 \text{ in}$$

The nominal strength for bolt bearing, R_n, is

$$R_n = \min \begin{cases} 1.2 L_c t_w F_u = (1.2)(2.19 \text{ in}) \\ \qquad \times (0.25 \text{ in})\left(65 \ \dfrac{\text{kips}}{\text{in}^2}\right) \\ \qquad = 42.7 \text{ kips} \\[6pt] 2.4 d t_w F_u = (2.4)\left(\dfrac{3}{4} \text{ in}\right)(0.25 \text{ in})\left(65 \ \dfrac{\text{kips}}{\text{in}^2}\right) \\ \qquad = 29.3 \text{ kips} \quad [\text{controls}] \end{cases}$$

The available connection strength, nR_n/Ω, for bolt bearing is

$$\frac{nR_n}{\Omega} = \frac{(3 \text{ bolts})(29.3 \text{ kips})}{2.00}$$

$$= 44.0 \text{ kips} > 26.3 \text{ kips} \quad \left[\begin{array}{l}\text{bolt bearing on the} \\ \text{beam web is OK}\end{array}\right]$$

> Use a PL $^5\!/_{16}$ in × $4\frac{1}{2}$ in × $8\frac{1}{2}$ in with three A325N, $\frac{3}{4}$ in diameter bolts and $\frac{1}{4}$ in fillet welds.

Refer to AISC *Steel Construction Manual* Fig. 10-11 within the Single-Plate Connections section of Chap. 10 for dimensional limitations on single-plate connections.

The connection sketch is shown.

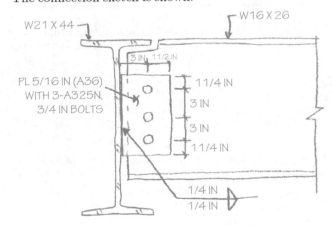

42. (c) The design axial compression load, P_r, and the design flexural moment, M_r, are determined using the provided shear and moment diagrams for unit reactions of beam M1. Consider P-δ effects by using the approximate second-order analysis procedure of AISC 360 App. 8. Per the problem statement, ignore lateral translation of the frame. So, $M_{lt} = P_{lt} = 0$, and Eq. A-8-1 and Eq. A-8-2 simplify to $M_r = B_1 M_{nt,\mathrm{top}}$ and $P_r = P_{nt}$, respectively. M_{nt} and P_{nt} are the first-order moment and axial force, respectively, assuming there is no translation of the frame.

Scale the unit shear and moment values by the beam reaction calculated in part 42(b) ($R_a = 26.3$ kips) to determine the first-order axial compression load, P_{nt}, and the first-order flexural moment, M_{nt}.

The first-order axial compression load, P_{nt}, is the sum of the reactions from beams M2 and M3.

The contribution from beam M2 is determined from the given loads as

$$P_{nt,\mathrm{M2}} = (\text{no. of floors})\left(\left(w_D + w_L\right)\left(\frac{L}{2}\right)\right)$$

$$= (2 \text{ floors})\left(\left(\frac{375 \,\frac{\mathrm{lbf}}{\mathrm{ft}}}{1000 \,\frac{\mathrm{lbf}}{\mathrm{kip}}} + \frac{500 \,\frac{\mathrm{lbf}}{\mathrm{ft}}}{1000 \,\frac{\mathrm{lbf}}{\mathrm{kip}}}\right)\left(\frac{30 \text{ ft}}{2}\right)\right)$$

$$= 26.25 \text{ kips}$$

The contribution from beam M3 is determined by scaling the shear forces from problem illustration II by the beam reaction determined in part 42(b).

$$P_{nt,\mathrm{M3}} = (26.3)(0.84 \text{ kip} + 0.87 \text{ kip}) = 45.0 \text{ kips}$$

The total design axial compression load is

$$\boxed{\begin{aligned} P_r &= P_{nt,\mathrm{M2}} + P_{nt,\mathrm{M3}} = 26.3 \text{ kips} + 45.0 \text{ kips} \\ &= 71.3 \text{ kips} \end{aligned}}$$

The first-order flexural moment, M_{nt}, is determined by scaling the moment from problem illustration III by the beam reaction determined in part 42(b).

The first-order flexural moment at the top of the column is

$$M_{nt,\mathrm{top}} = (26.3)(0.86 \text{ ft-kip}) = 22.6 \text{ ft-kips}$$

The design flexural moment at the bottom of the column is

$$\boxed{M_{nt,\mathrm{bot}} = M_{rx,\mathrm{bot}} = 0 \text{ ft-kips}}$$

Per AISC 360 App. 8 Sec. A8.2.1, the first-order moment must be amplified by B_1 to account for second-order effects caused by displacements between brace points (P-δ effects). From AISC 360 Eq. A8-4, coefficient C_m is

$$C_m = 0.6 - 0.4\left(\frac{M_1}{M_2}\right) = 0.6 - 0.4\left(\frac{M_{nt,\mathrm{bot}}}{M_{nt,\mathrm{top}}}\right)$$

$$= 0.6 - (0.4)\left(\frac{0 \text{ ft-kips}}{22.6 \text{ ft-kips}}\right)$$

$$= 0.6$$

The elastic critical buckling resistance of the column, P_{e1}, is calculated with respect to the plane of bending. The column is bent about the strong axis, so use I_x. From AISC *Steel Construction Manual* Table 1-1, for a W10 × 33 column, $I_x = 171$ in^4.

Using AISC 360 Eq. A-8-5, P_{e1} is

$$P_{e1} = \frac{\pi^2 EI}{(K_1 L)^2} = \frac{\pi^2\left(29{,}000 \,\frac{\mathrm{kips}}{\mathrm{in}^2}\right)(171 \text{ in}^4)}{\left((1.0)(15 \text{ ft})\left(12 \,\frac{\mathrm{in}}{\mathrm{ft}}\right)\right)^2} = 1511 \text{ kips}$$

From AISC 360 App. 8 Sec. A8.2.1, $\alpha = 1.0$ for LRFD. Using AISC 360 Eq. A-8-3, amplifier B_1 is

$$B_1 = \frac{C_m}{1 - \alpha\left(\dfrac{P_r}{P_{e1}}\right)} = \frac{0.6}{1 - (1.6)\left(\dfrac{71.2 \text{ kips}}{1511 \text{ kips}}\right)}$$

$$= 0.649 < 1.0$$

Therefore, $B_1 = 1.0$, and the design flexural moment at the top of the column is

$$M_r = B_1 M_{nt,\mathrm{top}} = (1.0)(22.6 \text{ ft-kips}) = 22.6 \text{ ft-kips}$$

42. (d) AISC 360 Comm. C-A-7.2 provides the procedure for determining the effective length factor, K, using alignment charts.

From AISC 360 Comm. C-A-7.2, for a pinned column end not rigidly connected to a footing or foundation, the alignment chart factor at the base of the column is

$$G_A = 10$$

From AISC 360 Comm. C-A-7.2, G_B is the sum of EI/L of rigidly connected columns framing into a joint divided by the sum of EI/L of rigidly connected girders framing into the joint. At the joint between the first and second floor, there are two columns framing into the joint (above and below the floor level) and one rigid girder framing into the joint (beam M3).

From AISC *Steel Construction Manual* Table 1-1, the moment of inertia of the $W10 \times 33$ column, I_c, is 171 in^4, and the moment of inertia of the $W21 \times 44$ girder, I_g, is 843 in^4. The unsupported lengths of the columns, L_c, and the girder, L_g, are determined from the problem illustrations.

The modulus of elasticity, E, is the same for all members and can be omitted.

The alignment chart factor at the top of the first floor column, G_B, is then

$$G_B = \frac{\Sigma\left(\dfrac{I}{L}\right)_c}{\Sigma\left(\dfrac{I}{L}\right)_g} = \frac{\dfrac{171\ \text{in}^4}{15\ \text{ft}} + \dfrac{171\ \text{in}^4}{12\ \text{ft}}}{\dfrac{843\ \text{in}^4}{25\ \text{ft}}} = 0.761$$

Use the alignment chart for sidesway uninhibited moment frames (see AISC 360 Comm. Fig. C-A-7.2) to determine the unbraced length, K_x.

For $G_A = 10$ and $G_B = 0.761$,

$$\boxed{K_x = 1.85}$$

Find the controlling value for KL/r. From AISC *Steel Construction Manual* Table 1-1, for a $W10 \times 33$ column, $r_x = 4.19$ in and $r_y = 1.94$ in. The controlling value for KL/r is the maximum of

$$\frac{KL}{r} = \max \begin{cases} \dfrac{K_x L_x}{r_x} = \dfrac{(1.85)(15\ \text{ft})\left(12\ \dfrac{\text{in}}{\text{ft}}\right)}{4.19\ \text{in}} \\ \qquad\quad = 79.5 \\ \dfrac{K_y L_y}{r_y} = \dfrac{(1.0)(15\ \text{ft})\left(12\ \dfrac{\text{in}}{\text{ft}}\right)}{1.94\ \text{in}} \\ \qquad\quad = 92.8 \quad [\text{controls}] \end{cases}$$

From AISC *Steel Construction Manual* Table 4-1, for $K_y L_y = 15$ ft, the available axial compression strength is

$$P_c = \frac{P_n}{\Omega_c} = 155\ \text{kips}$$

Find the available moment strength.

Because the flanges are laterally braced at the footing and at the first floor,

$$L_b = 15\ \text{ft}$$

From AISC *Steel Construction Manual* Table 3-2, $L_p = 6.85$ and $L_r = 21.8$ for a $W10 \times 33$ member. Therefore, $L_p < L_b \le L_r$, and the available moment is given by AISC 360 Eq. F2-2.

$$\frac{M_n}{\Omega_b} = \frac{C_b\left[M_p - (M_p - 0.7F_y S_x)\left(\dfrac{L_b - L_p}{L_r - L_p}\right)\right]}{\Omega_b} \le \frac{M_p}{\Omega_b}$$

From AISC *Steel Construction Manual* Table 3-2, for a $W10 \times 33$ column,

$$\frac{M_p}{\Omega_b} = 96.8\ \text{ft-kips}$$

From AISC 360 Sec. F1, when one end moment equals zero,

$$C_b = 1.67$$

From AISC *Steel Construction Manual* Table 3-10, for a $W10 \times 33$ with an unbraced length of 15 ft,

$$\frac{M_n'}{\Omega_b} = 77.5\ \text{ft-kips}$$

For the moment strengths from AISC *Steel Construction Manual* Table 3-10, assume $C_b = 1.0$. Therefore, in accordance with AISC 360 Eq. F2-2, the moment strength can be multiplied by C_b, but cannot exceed M_p/Ω_b.

$$\frac{M_n}{\Omega_b} = C_b\left(\frac{M_n'}{\Omega_b}\right) = (1.67)(77.5\ \text{ft-kips})$$

$$= 129\ \text{ft-kips} > M_p/\Omega_b \quad [M_p/\Omega_b\ \text{controls}]$$

The available strong-axis flexural strength, M_{cx}, is

$$M_{cx} = \frac{M_p}{\Omega_b} = 96.8\ \text{ft-kips}$$

Using the value for P_r found in part 42(c), find the value of P_r/P_c to determine the applicable interaction equation.

$$\frac{P_r}{P_c} = \frac{71.2\ \text{kips}}{155\ \text{kips}} = 0.46 > 0.2$$

The applicable interaction equation is AISC 360 Eq. H1-1a. Use the value for M_{rx} at the top of the column from part 42(c).

$$\frac{P_r}{P_c} + \left(\frac{8}{9}\right)\left(\frac{M_{rx}}{M_{cx}}\right) \le 1.0$$

$$= \frac{71.2 \text{ kips}}{155 \text{ kips}} + \left(\frac{8}{9}\right)\left(\frac{22.6 \text{ ft-kips}}{96.8 \text{ ft-kips}}\right)$$

$$= \boxed{0.67 \le 1.0 \quad [\text{OK}]}$$

Therefore, the $W10 \times 33$ column is adequate for combined flexure and compression under dead and live loads.

42. LRFD Solution

42. (a) The procedure for analyzing composite flexural members is given by AISC 360 Sec. I3.

From AISC 360 Sec. I3-1a for an interior beam, the effective slab width is the lesser of one-quarter of the span or the beam spacing.

$$b = \min \begin{cases} \dfrac{L}{4} = \left(\dfrac{30 \text{ ft}}{4}\right)\left(12 \dfrac{\text{in}}{\text{ft}}\right) = 90 \text{ in} \quad [\text{controls}] \\ s = \dfrac{B}{3} = \left(\dfrac{25 \text{ ft}}{3}\right)\left(12 \dfrac{\text{in}}{\text{ft}}\right) = 100 \text{ in} \end{cases}$$

For the case of the metal deck spanning perpendicular to the beam span, only the concrete above the metal deck will be effective. Therefore, the area of concrete slab within the effective slab width is

$$A_c = bt_s = (90 \text{ in})(3 \text{ in}) = 270 \text{ in}^2$$

From AISC *Steel Construction Manual* Table 1-1, the area, A_s, of a $W16 \times 26$ beam is

$$A_s = 7.68 \text{ in}^2$$

From AISC 360 Sec. I3-2d(1), the total horizontal shear, V', for a fully composite beam is the lesser of AISC 360 Eq. I3-1a or Eq. I3-1b.

$$V' = \min \begin{cases} 0.85f_c'A_c = (0.85)\left(4 \dfrac{\text{kips}}{\text{in}^2}\right)(270 \text{ in}^2) \\ \qquad = 918 \text{ kips} \\ F_yA_s = \left(50 \dfrac{\text{kips}}{\text{in}^2}\right)(7.68 \text{ in}^2) \\ \qquad = 384 \text{ kips} \quad [\text{controls}] \end{cases}$$

From AISC *Steel Construction Manual* Table 3-21, for a perpendicular deck and one $3/4$ in diameter strong stud per rib with 4 kips/in^2 normal-weight concrete, the nominal stud shear strength is

$$Q_n = 21.5 \text{ kips/stud}$$

To provide complete shear connection and full composite action, the required number of connectors on either side of the point of maximum moment is $n = V'/Q_n$.

For a simply supported beam, V'/Q_n must be multiplied by two to get the total number of required shear studs along the length of the beam. The required number of shear studs for full composite action is

$$2n = \frac{2V'}{Q_n} = \frac{(2)(384 \text{ kips})}{21.5 \dfrac{\text{kips}}{\text{stud}}} = \boxed{35.7 \text{ studs} \quad (36 \text{ studs})}$$

Check flexural strength.

Considering only the provided dead and live loads, from IBC Eq. 16-2, the distributed beam load, w_u, is

$$w_u = 1.2w_D + 1.6w_L$$

$$= \frac{(1.2)\left(750 \dfrac{\text{lbf}}{\text{ft}}\right) + (1.6)\left(1000 \dfrac{\text{lbf}}{\text{ft}}\right)}{1000 \dfrac{\text{lbf}}{\text{kip}}}$$

$$= 2.5 \text{ kips/ft}$$

For a simply supported beam, the required flexural strength, M_r, is

$$M_r = \frac{w_uL^2}{8} = \frac{\left(2.5 \dfrac{\text{kips}}{\text{ft}}\right)(30 \text{ ft})^2}{8}$$

$$= 281 \text{ ft-kips}$$

Following the procedure from AISC *Steel Construction Manual*, Part 3, the depth of the compression block, a, is

$$a = \frac{F_yA_s}{0.85f_c'b} = \frac{\left(50 \dfrac{\text{kips}}{\text{in}^2}\right)(7.68 \text{ in}^2)}{(0.85)\left(4 \dfrac{\text{kips}}{\text{in}^2}\right)(90 \text{ in})}$$

$$= 1.25 \text{ in}$$

The compression block is within the slab thickness ($a < Y_{\text{con}}$), so the distance from the top flange of the composite steel beam to the plastic neutral axis is $Y1 = 0$ in per AISC *Steel Construction Manual* Table 3-19.

The distance, $Y2$, from the concrete flange force to the steel beam top flange is

$$Y2 = Y_{con} - \frac{a}{2} = 6 \text{ in} - \frac{1.25 \text{ in}}{2}$$
$$= 5.38 \text{ in}$$

Interpolating from AISC *Steel Construction Manual* Table 3-19, for a W16 × 26 beam,

$$\phi_b M_p = \boxed{381 \text{ ft-kips} > 281 \text{ ft-kips} \quad [\text{OK}]}$$

> Therefore, the composite beam is adequate in flexure.

Check deflection.

The allowable deflection limit, $\Delta_{allowable}$, is

$$\Delta_{allowable} = \frac{L}{360} = \frac{(30 \text{ ft})\left(12 \dfrac{\text{in}}{\text{ft}}\right)}{360}$$
$$= 1 \text{ in}$$

Use the lower bound elastic moment of inertia, I_{LB}, to calculate the live load deflection. Interpolating from AISC *Steel Construction Manual* Table 3-20, for a W16 × 26 beam, $I_{LB} = 972 \text{ in}^4$.

$$\Delta_L = \frac{5 w_L L^4}{384 EI}$$

$$= \frac{(5)\left(\dfrac{1000 \dfrac{\text{lbf}}{\text{ft}}}{\left(1000 \dfrac{\text{lbf}}{\text{kip}}\right)\left(12 \dfrac{\text{in}}{\text{ft}}\right)}\right)\left((30 \text{ ft})\left(12 \dfrac{\text{in}}{\text{ft}}\right)\right)^4}{(384)\left(29{,}000 \dfrac{\text{kips}}{\text{in}^2}\right)(972 \text{ in}^4)}$$

$$= \boxed{0.65 \text{ in} < 1 \text{ in} \quad [\text{OK}]}$$

> The composite beam meets the deflection criteria.

42. (b) The beam reaction, R_u, is

$$R_u = \frac{w_u L}{2} = \frac{\left(2.5 \dfrac{\text{kips}}{\text{ft}}\right)(30 \text{ ft})}{2} = 37.5 \text{ kips}$$

Available strengths for single-plate connections are given in AISC *Steel Construction Manual* Table 10-10a. Using the table, the most efficient connection design with an available strength, ϕR_n, greater than the beam

reaction, R_u, is a connection with three A325 bolts ($n = 3$, $L = 8.5$ in) and a plate thickness of $t_p = \frac{1}{4}$ in. A325 bolts are in bolt group A. Conservatively assume condition of threads included (N) in the faying surface. From the bottom of AISC *Steel Construction Manual* Table 10-10a, the weld size for the plate to girder connection is $\frac{3}{16}$ in.

From AISC *Steel Construction Manual* Table 10-10a, the available connection strength is

$$\phi R_n = 38.3 \text{ kips} > 37.5 \text{ kips} \quad [\text{OK}]$$

From the AISC *Steel Construction Manual* description for Table 10-10a, the tabulated strengths consider the limit-states of bolt shear, bolt bearing on the plate, shear yielding of the plate, shear rupture of the plate, block shear rupture of the plate, and weld shear. However, AISC *Steel Construction Manual* Table 10-10a does not include a check for bolt bearing on the beam web. From AISC *Steel Construction Manual* Table 1-1, the beam web thickness, t_w, is 0.25 in, which is the same as the plate thickness. Therefore, bolt bearing on the beam does not need to be checked per AISC 360 Sec. J3-10.

> Use a PL $\frac{1}{4}$ in × $4\frac{1}{2}$ in × $8\frac{1}{2}$ in with three A325N, $\frac{3}{4}$ in diameter bolts and $\frac{3}{16}$ in fillet welds.

Refer to AISC *Steel Construction Manual* Fig. 10-11 within the Single-Plate Connections section of Chap. 10 for dimensional limitations on single-plate connections.

The connection sketch is shown.

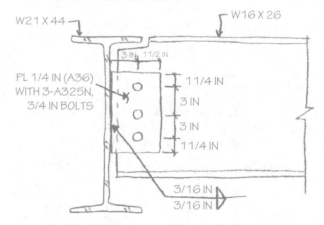

42. (c) The design axial compression load, P_r, and the design flexural moment, M_r, are determined using the provided shear and moment diagrams for unit reactions of beam M1. Consider P-δ effects by using the approximate second-order analysis procedure of AISC 360 App. 8. Per the problem statement, ignore lateral translation of the frame. So, $M_{lt} = P_{lt} = 0$, and Eq. A-8-1 and Eq. A-8-2 simplify to $M_r = B_1 M_{nt,top}$ and $P_r = P_{nt}$, respectively. M_{nt} and P_{nt} are the first-order moment and axial force, respectively, assuming there is no translation of the frame.

Scale the unit shear and moment values by the beam reaction, calculated in part 42(b) ($R_u = 37.5$ kips), to determine the first-order axial compression load, P_{nt}, and the first-order flexural moment, M_{nt}.

The first-order axial compression load, P_{nt}, is the sum of the reactions from beams M2 and M3.

The contribution from beam M2 is determined from the given loads as

$$P_{nt,M2} = (\text{no. of floors})\left[(1.2w_D + 1.6w_L)\left(\frac{L}{2}\right)\right]$$

$$= (2 \text{ floors})\left[\left(\frac{(1.2)\left(375 \frac{\text{lbf}}{\text{ft}}\right)}{1000 \frac{\text{lbf}}{\text{kip}}} + \frac{(1.6)\left(500 \frac{\text{lbf}}{\text{ft}}\right)}{1000 \frac{\text{lbf}}{\text{kip}}}\right)\left(\frac{30 \text{ ft}}{2}\right)\right]$$

$$= 37.5 \text{ kips}$$

The contribution from beam M3 is determined by scaling the shear forces from problem illustration II by the beam reaction determined in part 42(b).

$$P_{nt,M3} = (37.5)(0.84 \text{ kip} + 0.87 \text{ kip}) = 64.1 \text{ kips}$$

The total design axial compression load is

$$\boxed{\begin{aligned} P_r &= P_{nt,M2} + P_{nt,M3} = 37.5 \text{ kips} + 64.1 \text{ kips} \\ &= 102 \text{ kips} \end{aligned}}$$

The first-order flexural moment, M_{nt}, is determined by scaling the moment from problem illustration III by the beam reaction determined in part 42(b).

The first-order flexural moment at the top of the column is

$$M_{nt,\text{top}} = (37.5)(0.86 \text{ ft-kip}) = 32.3 \text{ ft-kips}$$

The design flexural moment at the bottom of the column is

$$\boxed{M_{nt,\text{bot}} = M_{rx,\text{bot}} = 0 \text{ ft-kips}}$$

Per AISC 360 App. 8 Sec. A8.2.1, the first-order moment must be amplified by B_1 to account for second-order effects caused by displacements between brace points (P-δ effects). From AISC 360 Eq. A-8-4, coefficient C_m is

$$\begin{aligned} C_m &= 0.6 - 0.4\left(\frac{M_1}{M_2}\right) = 0.6 - 0.4\left(\frac{M_{nt,\text{bot}}}{M_{nt,\text{top}}}\right) \\ &= 0.6 - (0.4)\left(\frac{0 \text{ ft-kips}}{22.6 \text{ ft-kips}}\right) \\ &= 0.6 \end{aligned}$$

The elastic critical buckling resistance of the column, P_{e1}, is calculated with respect to the plane of bending. The column is bent about the strong axis, so use I_x. From AISC *Steel Construction Manual* Table 1-1, for a W10 × 33 column, $I_x = 171 \text{ in}^4$.

Using AISC 360 Eq. A-8-5, P_{e1} is

$$P_{e1} = \frac{\pi^2 EI_x}{(K_1 L)^2} = \frac{\pi^2\left(29{,}000 \frac{\text{kips}}{\text{in}^2}\right)(171 \text{ in}^4)}{\left((1.0)(15 \text{ ft})\left(12 \frac{\text{in}}{\text{ft}}\right)\right)^2} = 1511 \text{ kips}$$

From AISC 360 App. 8, Sec. A 8.2.1, $\alpha = 1.0$ for LRFD. Using AISC 360 Eq. A-8-3, amplifier B_1 is

$$\begin{aligned} B_1 &= \frac{C_m}{1 - \alpha\left(\frac{P_r}{P_{e1}}\right)} = \frac{0.6}{1 - (1.0)\left(\frac{102 \text{ kips}}{1511 \text{ kips}}\right)} \\ &= 0.643 < 1.0 \end{aligned}$$

Therefore, $B_1 = 1.0$, and the design flexural moment at the top of the column is

$$M_r = B_1 M_{nt,\text{top}} = (1.0)(32.3 \text{ ft-kips}) = 32.3 \text{ ft-kips}$$

42. (d) AISC 360 Comm. C-A-7.2 provides the procedure for determining the effective length factor, K, using alignment charts.

From AISC 360 Comm. C-A-7.2, for a pinned column end not rigidly connected to a footing or foundation, the alignment chart factor at the base of the column is

$$G_A = 10$$

From AISC 360 Comm. C-A-7.2, G_B is the sum of EI/L of rigidly connected columns framing into a joint divided by the sum of EI/L of rigidly connected girders framing into the joint. At the joint between the first and second floor, there are two columns framing into the joint (above and below the floor level) and one rigid girder framing into the joint (beam M3).

From AISC *Steel Construction Manual* Table 1-1, the moment of inertia of the W10 × 33 column, I_c, is 171 in⁴, and the moment of inertia of the W21 × 44 girder, I_g, is 843 in⁴. The unsupported lengths of the columns, L_c, and the girder, L_g, are determined from the problem illustrations.

The modulus of elasticity, E, is the same for all members and can be omitted.

The alignment chart factor at the top of the first floor column, G_B, is then

$$G_B = \frac{\sum \left(\dfrac{I}{L}\right)_c}{\sum \left(\dfrac{I}{L}\right)_g} = \frac{\dfrac{171 \text{ in}^4}{15 \text{ ft}} + \dfrac{171 \text{ in}^4}{12 \text{ ft}}}{\dfrac{843 \text{ in}^4}{25 \text{ ft}}} = 0.761$$

Use the alignment chart for sidesway uninhibited moment frames (see AISC 360 Comm. Fig. C-A-7.2) to determine the unbraced length, K_x.

For $G_A = 10$ and $G_B = 0.761$,

$$\boxed{K_x = 1.85}$$

Find the controlling value for KL/r. From AISC *Steel Construction Manual* Table 1-1, for a W10 × 33 column, $r_x = 4.19$ in and $r_y = 1.94$ in. The controlling value for KL/r is the maximum of

$$\frac{KL}{r} = \max \begin{cases} \dfrac{K_x L_x}{r_x} = \dfrac{(1.85)(15 \text{ ft})\left(12 \dfrac{\text{in}}{\text{ft}}\right)}{4.19 \text{ in}} \\ \qquad = 79.5 \\ \dfrac{K_y L_y}{r_y} = \dfrac{(1.0)(15 \text{ ft})\left(12 \dfrac{\text{in}}{\text{ft}}\right)}{1.94 \text{ in}} \\ \qquad = 92.8 \quad \text{[controls]} \end{cases}$$

From AISC *Steel Construction Manual* Table 4-1, for $K_y L_y = 15$ ft, the available axial compression strength is

$$P_c = \phi_c P_n = 233 \text{ kips}$$

Find the available moment strength.

Because the flanges are laterally braced at the footing and at the first floor,

$$L_b = 15 \text{ ft}$$

From AISC *Steel Construction Manual* Table 3-2, $L_p = 6.85$ and $L_r = 21.8$ for a W10 × 33 member. Therefore,

$L_p < L_b \le L_r$ and the available moment is given by AISC 360 Eq. F2-2.

$$\phi_b M_n = \phi_b C_b \left[M_p - (M_p - 0.7 F_y S_x)\left(\frac{L_b - L_p}{L_r - L_p}\right) \right]$$
$$\le \phi_b M_p$$

From AISC *Steel Construction Manual* Table 3-2, for a W10 × 33 column,

$$\phi_b M_p = 146 \text{ ft-kips}$$

From AISC 360 Sec. F1, when one end moment equals zero,

$$C_b = 1.67$$

From AISC *Steel Construction Manual* Table 3-10, for a W10 × 33 with an unbraced length of 15 ft,

$$\phi_b M_n' = 116 \text{ ft-kips}$$

For the moment strengths from AISC *Steel Construction Manual* Table 3-10, assume $C_b = 1.0$. Therefore, in accordance with AISC 360 Eq. F2-2, the moment strength can be multiplied by C_b, but cannot exceed $\phi_b M_p$.

$$\phi_b M_n = C_b \phi_b M_n' = (1.67)(116 \text{ ft-kips})$$
$$= 194 \text{ ft-kips} > \phi_b M_p \quad [\phi_b M_p \text{ controls}]$$

Therefore, the available strong-axis flexural strength, M_{cx}, is

$$M_{cx} = \phi_b M_p = 146 \text{ ft-kips}$$

Using the value for P_r found in part 42(c), find the value of P_r/P_c to determine the applicable interaction equation.

$$\frac{P_r}{P_c} = \frac{102 \text{ kips}}{233 \text{ kips}} = 0.44 > 0.2$$

Therefore, the applicable interaction equation is AISC 360 Eq. H1-1a. Use the value for M_{rx} at the top of the column from part 42(c).

$$\frac{P_r}{P_c} + \left(\frac{8}{9}\right)\left(\frac{M_{rx}}{M_{cx}}\right) \le 1.0$$

$$= \frac{102 \text{ kips}}{233 \text{ kips}} + \left(\frac{8}{9}\right)\left(\frac{32.3 \text{ ft-kips}}{146 \text{ ft-kips}}\right)$$

$$= \boxed{0.63 \le 1.0 \quad \text{[OK]}}$$

Therefore, the W10 × 33 column is adequate for combined flexure and compression under dead and live loads.

43. (a) The factor of safety against sliding, $(FS)_{sliding}$, for a cantilever retaining wall is the sum of the horizontal resisting (or restoring) forces divided by the sum of the horizontal driving forces. Load factors are traditionally not used for stability calculations. To be considered stable, the factor of safety against sliding should be at least 1.5.

The horizontal driving forces are the lateral earth pressures resulting from the weight of the soil and the surcharge from the concrete slab. These driving forces are calculated using the coefficient of active soil pressure because the retaining wall is not restrained from movement at the top.

The horizontal resisting forces are the passive soil resistance at the toe and the frictional resistance of the base due to the total vertical load. Typically, soil above the toe is ignored (in this problem, there is no soil) for stability calculation because this soil may be excavated during the lifespan of the structure. Transient surcharge loads are usually omitted in the calculation of the frictional resistance; however, in this problem the surcharge load from the concrete slab is included because it is permanent.

Calculate the horizontal driving forces.

The net horizontal force due the surcharge load from the slab, F_{sur}, is

$$
\begin{aligned}
F_{sur} &= K_a \gamma_c t_{slab}(h_{soil} + t_{base}) \\
&= (0.42)\left(150 \ \frac{lbf}{ft^3}\right)(1 \ ft)(5 \ ft + 1.25 \ ft) \\
&= 394 \ lbf/ft
\end{aligned}
$$

The net horizontal force due to soil loads, F_{soil}, is

$$
\begin{aligned}
F_{soil} &= \tfrac{1}{2}\gamma_s K_a(h_{soil} + t_{base})^2 \\
&= \left(\frac{1}{2}\right)\left(115 \ \frac{lbf}{ft^3}\right)(0.42)(5 \ ft + 1.25 \ ft)^2 \\
&= 943 \ lbf/ft
\end{aligned}
$$

The total horizontal load, $F_{h,tot}$, is

$$
\begin{aligned}
F_{h,tot} &= F_{sur} + F_{soil} = 394 \ \frac{lbf}{ft} + 943 \ \frac{lbf}{ft} \\
&= 1340 \ lbf/ft
\end{aligned}
$$

The vertical forces acting on the foundation include the weight of the CMU wall stem, concrete base, and soil above the heel, as well as the surcharge from the concrete slab. The total vertical load is needed to determine the total sliding resistance, $F_{res,tot}$.

The vertical force from the weight of the CMU wall stem, W_{wall}, is

$$
\begin{aligned}
W_{wall} &= w_m h_{wall} = \left(105 \ \frac{lbf}{ft^2}\right)(5 \ ft + 1 \ ft + 3 \ ft) \\
&= 945 \ lbf/ft
\end{aligned}
$$

The vertical force from the weight of the base, W_{base}, is

$$
\begin{aligned}
W_{base} &= \gamma_c t_{base} l_{base} = \left(150 \ \frac{lbf}{ft^3}\right)(1.25 \ ft)(6.8 \ ft) \\
&= 1275 \ lbf/ft
\end{aligned}
$$

The vertical force from the weight of the slab (surcharge), W_{sur}, is

$$
\begin{aligned}
W_{sur} &= \gamma_c t_{slab} l_{heel} = \left(150 \ \frac{lbf}{ft^3}\right)(1 \ ft)(4 \ ft) \\
&= 600 \ lbf/ft
\end{aligned}
$$

The vertical force from the weight of the soil above the heel, W_{soil}, is

$$
\begin{aligned}
W_{soil} &= \gamma_s h_{soil} l_{heel} = \left(115 \ \frac{lbf}{ft^3}\right)(5 \ ft)(4 \ ft) \\
&= 2300 \ lbf/ft
\end{aligned}
$$

The total vertical load, $W_{v,tot}$, is

$$
\begin{aligned}
W_{v,tot} &= W_{wall} + W_{base} + W_{sur} + W_{soil} \\
&= 945 \ \frac{lbf}{ft} + 1275 \ \frac{lbf}{ft} + 600 \ \frac{lbf}{ft} + 2300 \ \frac{lbf}{ft} \\
&= 5120 \ lbf/ft
\end{aligned}
$$

The restorative force due to friction, F_{fric}, is the total vertical dead load multiplied by the coefficient of static friction.

$$
F_{fric} = W_{tot}\mu = \left(5120 \ \frac{lbf}{ft}\right)(0.35) = 1790 \ lbf/ft
$$

The restorative force due to passive soil resistance on the front edge of the concrete base, F_p, is

$$
\begin{aligned}
F_p &= \tfrac{1}{2}\gamma_s K_p t_{base}^2 = \left(\frac{1}{2}\right)\left(115 \ \frac{lbf}{ft^3}\right)(2.39)(1.25 \ ft)^2 \\
&= 215 \ lbf/ft
\end{aligned}
$$

The total sliding resistance, $F_{\text{res,tot}}$, is

$$F_{\text{res,tot}} = F_p + F_{\text{fric}} = 215 \ \frac{\text{lbf}}{\text{ft}} + 1790 \ \frac{\text{lbf}}{\text{ft}}$$

$$= 2010 \ \text{lbf/ft}$$

The factor of safety against sliding is

$$(\text{FS})_{\text{sliding}} = \frac{F_{\text{res,tot}}}{F_{h,\text{tot}}} = \frac{2010 \ \dfrac{\text{lbf}}{\text{ft}}}{1340 \ \dfrac{\text{lbf}}{\text{ft}}} = \boxed{1.50}$$

Sketch the horizontal forces acting on the wall.

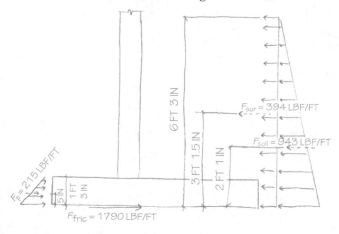

43. (b) From IBC Eq. 16-9, use a load factor of 1.0 for the dead loads, D, and lateral earth pressures, H, acting on the wall.

Calculate the forces acting on a 1 ft width of wall.

The total compressive force per foot width of wall, P, at the base of the stem is equal to the weight of the wall calculated in part 43(a) as $P = W_{\text{wall}} = 945 \ \text{lbf}$.

Considering only the height of the soil and not including the thickness of the concrete base, the shear force per foot width of wall, V, at the base of the CMU stem is the sum of the horizontal forces due to slab surcharge and soil.

The horizontal force due to the slab surcharge, F'_{sur}, is

$$F'_{\text{sur}} = K_a \gamma_c t_{\text{slab}} h_{\text{soil}}(1 \ \text{ft})$$

$$= (0.42)\left(150 \ \frac{\text{lbf}}{\text{ft}^3}\right)(1 \ \text{ft})(5 \ \text{ft})(1 \ \text{ft})$$

$$= 315 \ \text{lbf}$$

The horizontal force due to the soil, F'_{soil}, is

$$F'_{\text{soil}} = \tfrac{1}{2}\gamma_s K_a h^2_{\text{soil}}(1 \ \text{ft}) = \left(\frac{1}{2}\right)\left(115 \ \frac{\text{lbf}}{\text{ft}^3}\right)(0.42)(5 \ \text{ft})^2(1 \ \text{ft})$$

$$= 604 \ \text{lbf}$$

The shear force per foot width of wall at the base of the CMU stem is then

$$V = F'_{\text{sur}} + F'_{\text{soil}} = 315 \ \text{lbf} + 604 \ \text{lbf} = 919 \ \text{lbf}$$

The moment per foot width of wall, M, at the base of the CMU wall is calculated by multiplying the horizontal forces by the distance between the base of the stem and the line of action of the forces.

$$M = \tfrac{1}{2}h_{\text{soil}}F'_{\text{sur}} + \tfrac{1}{3}h_{\text{soil}}F'_{\text{soil}}$$

$$= \left(\frac{1}{2}\right)(5 \ \text{ft})(315 \ \text{lbf}) + \left(\frac{1}{3}\right)(5 \ \text{ft})(604 \ \text{lbf})$$

$$= 1790 \ \text{ft-lbf}$$

Check the compressive stress due to the axial loads.

The only axial load on the wall is the wall's self-weight. This load is relatively small.

> By inspection, the wall is adequate in compression due to axial loads alone.

Alternatively, the axial load capacity can be calculated as follows.

The calculated compression force, P, due to axial loads alone cannot exceed the allowable values given in TMS 402 Sec. 8.3.4.2.1. For a cantilever, the effective height of the wall is $2h_{\text{wall}}$.

$$\frac{h_{\text{wall}}}{r} = \frac{(2)(5 \ \text{ft} + 1 \ \text{ft} + 3 \ \text{ft})\left(12 \ \dfrac{\text{in}}{\text{ft}}\right)}{\dfrac{9.625 \ \text{in}}{\sqrt{12}}} = 77.7 < 99$$

For columns having an h_{wall}/r ratio less than 99, TMS 402 Eq. 8-21 applies. Ignoring the contribution from the steel, the allowable axial load, P_a, is

$$
\begin{aligned}
P_a &= (0.25f'_m A_n)\left[1 - \left(\frac{h_{wall}}{140r}\right)^2\right] \\
&= \left[(0.25)\left(1500\ \frac{lbf}{in^2}\right)(1\ ft)\left(12\ \frac{in}{ft}\right)(9.625\ in)\right] \\
&\quad \times \left[1 - \left(\frac{77.7}{140}\right)^2\right] \\
&= 30{,}000\ lbf > 945\ lbf \quad [OK]
\end{aligned}
$$

Therefore, the wall is adequate for compression due to axial loads alone.

Check the flexural stress. The procedure for determining the stresses in a reinforced masonry section is outlined in many structural engineering references.

From TMS 402 Sec. 4.2.2.1, the modulus of elasticity of steel reinforcement, E_s, is

$$E_s = 29{,}000{,}000\ lbf/in^2$$

From TMS 402 Sec. 4.2.2.2.1, the modulus of elasticity for concrete masonry, E_m, is

$$E_m = 900f'_m = (900)\left(1500\ \frac{lbf}{in^2}\right) = 1{,}350{,}000\ lbf/in^2$$

The modular ratio, n, is

$$
n = \frac{E_s}{E_m} = \frac{29{,}000{,}000\ \dfrac{lbf}{in^2}}{1{,}350{,}000\ \dfrac{lbf}{in^2}} = 21.5
$$

Considering a 1 ft width of wall and no. 5 bars (with a nominal area of 0.31 in^2 per ACI 318 App. A) at 24 in o.c., the area of steel, A_s, is

$$
A_s = A_{bar}\left(\frac{1\ ft\ wall\ width}{bar\ spacing}\right) = (0.31\ in^2)\left(\frac{1\ ft}{24\ in}\right)\left(12\ \frac{in}{ft}\right)
$$

$$= 0.155\ in^2$$

The reinforcement ratio, ρ, for a 1 ft width of wall is

$$
\rho = \frac{A_s}{b_w d} = \frac{0.155\ in^2}{(1\ ft)\left(12\ \dfrac{in}{ft}\right)(7\ in)} = 0.001845
$$

$$\rho n = (0.001845)(21.5) = 0.0396$$

The neutral axis depth factor, k, is

$$
\begin{aligned}
k &= \sqrt{2\rho n + (\rho n)^2} - \rho n \\
&= \sqrt{(2)(0.0396) + (0.0396)^2} - 0.0396 \\
&= 0.245
\end{aligned}
$$

The lever-arm factor, j, is

$$
j = 1 - \frac{k}{3} = 1 - \frac{0.245}{3} = 0.918
$$

The steel stress, f_s, is

$$
f_s = \frac{M}{A_s jd} = \frac{(1790\ ft\text{-}lbf)\left(12\ \dfrac{in}{ft}\right)}{(0.155\ in^2)(0.918)(7\ in)} = 21{,}600\ lbf/in^2
$$

From TMS 402 Sec. 8.3.3.1, the allowable steel tensile stress, F_s, for grade 60 reinforcement is 32,000 lbf/in^2.

$$\boxed{F_s = 32{,}000\ lbf/in^2 > 21{,}600\ lbf/in^2 \quad [OK]}$$

The stress in the masonry, f_b, due to the applied moment is

$$
f_b = \frac{2M}{jkb_w d^2} = \frac{(2)(1790\ ft\text{-}lbf)}{(0.918)(0.245)(1\ ft)(7\ in)^2} = 325\ lbf/in^2
$$

Using the value for P from part 43(b), the calculated compressive stress in the masonry due only to axial loads, f_a, is

$$
f_a = \frac{P}{b_w h} = \frac{945\ lbf}{(1\ ft)\left(12\ \dfrac{in}{ft}\right)(9.625\ in)} = 8.18\ lbf/in^2
$$

Directly combining the compressive stress due to flexure and direct axial loads (conservative), the total calculated stress in the masonry, f_m, is

$$
f_m = f_b + f_a = 325\ \frac{lbf}{in^2} + 8.18\ \frac{lbf}{in^2} = 333\ lbf/in^2
$$

From TMS 402 Sec. 8.3.4.2.2, the allowable masonry compressive stress in the masonry due to flexure, F_b, is

$$
F_b = 0.45f'_m = (0.45)\left(1500\ \frac{lbf}{in^2}\right)
$$

$$\boxed{= 675\ lbf/in^2 > 333\ lbf/in^2 \quad [OK]}$$

Therefore, the masonry wall is adequate for combined compression and flexure.

Check shear.

From TMS 402 Eq. 8-24, the shear stress in the masonry, f_v, is

$$f_v = \frac{V}{A_{nv}} = \frac{919 \text{ lbf}}{(1 \text{ ft})\left(12 \frac{\text{in}}{\text{ft}}\right)(9.625 \text{ in})} = 7.96 \text{ lbf/in}^2$$

Using $d = 7$ in instead of $d = 9.625$ in in the calculation of A_{nw} above is conservative and would also be considered correct.

Determine the masonry allowable shear stress in accordance with TMS 402 Sec. 8.3.5.1.3. Per the commentary, it is permissible to conservatively take $M/vd = 1.0$.

From TMS 402 Eq. 8-29, using f_a for $P/A_n = P/b_w h$,

$$\begin{aligned}
F_{vm} &= \frac{1}{2}\left[\left(4.0 - 1.75\left(\frac{M}{vd}\right)\right)\sqrt{f'_m}\right] + 0.25 f_a \\
&= \left(\frac{1}{2}\right)\left[\left(4.0 - (1.75)(1.0)\right)\sqrt{1500 \frac{\text{lbf}}{\text{in}^2}}\right] \\
&\quad + (0.25)\left(8.18 \frac{\text{lbf}}{\text{in}^2}\right) \\
&= 45.6 \text{ lbf/in}^2
\end{aligned}$$

From TMS 402 Eq. 8-27,

$$\begin{aligned}
F_v \leq 2\sqrt{f'_m}\gamma_g &= 2\sqrt{1500 \frac{\text{lbf}}{\text{in}^2}}\,(1.0) \\
&= 77.5 \text{ lbf/in}^2 \quad \text{[does not control]}
\end{aligned}$$

$$\boxed{F_v = 45.6 \text{ lbf/in}^2 > 7.96 \text{ lbf/in}^2 \quad \text{[OK]}}$$

Therefore, the masonry wall is adequate for shear.

43. (c) For the loads shown in problem illustration II, the concrete footing requires top bars for flexure at the heel and bottom bars for flexure at the toe.

For the top bars at the heel of the foundation, the concrete is exposed to earth but not cast against it. From ACI 318 Table 20.6.1.3.1, the minimum cover, c_c, for a no. 4 bar is $1\frac{1}{2}$ in. (Using a larger cover of 3 in is conservative but acceptable.) From ACI 318 App. A, the

nominal diameter of a no. 4 bar is 0.5 in. The effective depth, d, for flexure in the heel is

$$\begin{aligned}
d &= t_{\text{base}} - c_c - \frac{d_b}{2} = (1.25 \text{ ft})\left(12 \frac{\text{in}}{\text{ft}}\right) - 1.5 \text{ in} - \frac{0.5 \text{ in}}{2} \\
&= 13.25 \text{ in}
\end{aligned}$$

Check shear.

The critical section for shear is determined per ACI 318 Sec. 13.2.7.2. From ACI 318 Sec. 7.4.3.2, in order for the maximum shear, V_u, at the supports to be calculated at the critical section, conditions (a), (b), and (c) must be met.

ACI 318 Sec. 7.4.3.2 condition (a) requires that support reactions, in the direction of applied shear, introduce compression into the end regions of the member. Because the support reaction introduces tension into the region of the footing under the CMU stem, condition (a) is not met. Thus, the critical section for shear cannot be taken at a distance, d, from the face of the wall and must be taken at the face of the wall.

From the factored loads in problem illustration II, the slope, m, of the soil bearing pressure reaction is

$$m = \frac{q_{\text{toe}} - q_{\text{heel}}}{l_{\text{base}}} = \frac{1660 \frac{\text{lbf}}{\text{ft}^2} - 320 \frac{\text{lbf}}{\text{ft}^2}}{6.8 \text{ ft}} = 197 \text{ (lbf/ft}^2)/\text{ft}$$

The bearing pressure at the critical shear section (the back face of the wall), $q_{b,v}$, is

$$\begin{aligned}
q_{b,v} &= q_{\text{heel}} + m l_{\text{heel}} = 320 \frac{\text{lbf}}{\text{ft}^2} + \left(197 \frac{\frac{\text{lbf}}{\text{ft}^2}}{\text{ft}}\right)(4 \text{ ft}) \\
&= 1108 \text{ lbf/ft}^2 \quad (1110 \text{ lbf/ft}^2)
\end{aligned}$$

The factored shear, V_u, at the heel is

$$\begin{aligned}
V_u &= \left(w_{u,\text{heel}} - \frac{1}{2}(q_{b,v} + q_{\text{heel}})\right) l_{\text{heel}} \\
&= \frac{\left(1180 \frac{\text{lbf}}{\text{ft}^2} - \left(\frac{1}{2}\right)\left(1110 \frac{\text{lbf}}{\text{ft}^2} + 320 \frac{\text{lbf}}{\text{ft}^2}\right)\right)(4 \text{ ft})}{1000 \frac{\text{lbf}}{\text{kip}}} \\
&= 1.86 \text{ kips/ft}
\end{aligned}$$

From ACI 318 Table 21.2.1, $\phi = 0.75$ for shear.

From ACI 318 Table 19.2.4.2, $\lambda = 1.0$ for normal weight concrete.

ACI 318 Eq. 22.5.5.1 gives the shear strength provided by a 1 ft width of concrete, ϕV_c.

$$\phi V_c = \phi 2\lambda\sqrt{f'_c}\, b_w d$$

$$= \frac{(0.75)(2)(1.0)\sqrt{4000\,\dfrac{\text{lbf}}{\text{in}^2}}(1\text{ ft})\left(12\,\dfrac{\text{in}}{\text{ft}}\right)(13.25\text{ in})}{1000\,\dfrac{\text{lbf}}{\text{kip}}}$$

$$= \boxed{15.1\text{ kips/ft} > 1.86\text{ kips/ft} \quad [\text{OK}]}$$

From ACI 318 Sec. 7.6.3.1, no shear reinforcement is required.

$$\boxed{\text{Therefore, the footing is adequate in shear.}}$$

Check flexure.

From ACI 318 Table 13.2.7.1, the critical section for moment is halfway between the middle and edge of a masonry wall.

The bearing pressure at the critical section for moment, $q_{b,m}$, is

$$q_{b,m} = q_{\text{heel}} + m\left(l_{\text{heel}} + \frac{t_{\text{wall}}}{4}\right)$$

$$= 320\,\frac{\text{lbf}}{\text{ft}^2} + \left(197\,\frac{\dfrac{\text{lbf}}{\text{ft}^2}}{\text{ft}}\right)\left(4\text{ ft} + \frac{9.625\text{ in}}{\left(12\,\dfrac{\text{in}}{\text{ft}}\right)(4)}\right)$$

$$= 1150\text{ lbf/ft}^2$$

The critical factored moment at the heel, $M_{u,\text{heel}}$, (per foot width) is

$$M_{u,\text{heel}} = w_{u,\text{heel}}l_{\text{heel}}\left(\frac{l_{\text{heel}}}{2} + \frac{t_{\text{wall}}}{4}\right)$$

$$- \frac{(q_{b,m} + 2q_{\text{heel}})\left(l_{\text{heel}} + \dfrac{t_{\text{wall}}}{4}\right)^2}{6}$$

$$= \frac{\begin{array}{c}\left(1180\,\dfrac{\text{lbf}}{\text{ft}^2}\right)(4\text{ ft})\left(\dfrac{4\text{ ft}}{2} + \dfrac{9.625\text{ in}}{(4)\left(12\,\dfrac{\text{in}}{\text{ft}}\right)}\right) \\[10pt] \left(1150\,\dfrac{\text{lbf}}{\text{ft}^2} + (2)\left(320\,\dfrac{\text{lbf}}{\text{ft}^2}\right)\right) \\[10pt] \times\left(4\text{ ft} + \dfrac{9.625\text{ in}}{(4)\left(12\,\dfrac{\text{in}}{\text{ft}}\right)}\right)^2 \\[6pt] \dfrac{}{6}\end{array}}{1000\,\dfrac{\text{lbf}}{\text{kip}}}$$

$$= 5.12\text{ ft-kips/ft}$$

Check the adequacy of no. 4 bars (nominal area of 0.20 in^2 per ACI 318 App. A) at 6 in o.c. for flexure.

The area of the reinforcement provided, $A_{s,\text{prov}}$, in a 1 ft width of wall is

$$A_{s,\text{prov}} = A_{\text{bar}}\left(\frac{1\text{ ft wall width}}{\text{bar spacing}}\right)$$

$$= (0.20\text{ in}^2)\left(\frac{1\text{ ft}}{6\text{ in}}\right)\left(12\,\frac{\text{in}}{\text{ft}}\right)$$

$$= 0.40\text{ in}^2$$

The depth of the compression block, a, is

$$a = \frac{A_{s,\text{prov}}f_y}{0.85f'_c b_w} = \frac{(0.40\text{ in}^2)\left(60\,\dfrac{\text{kips}}{\text{in}^2}\right)}{(0.85)\left(\dfrac{4000\,\dfrac{\text{lbf}}{\text{in}^2}}{1000\,\dfrac{\text{lbf}}{\text{kip}}}\right)(1\text{ ft})\left(12\,\dfrac{\text{in}}{\text{ft}}\right)}$$

$$= 0.588\text{ in}$$

By inspection, the section is tension controlled.

Therefore, from ACI 318 Table 21.2.1, $\phi = 0.9$ and the flexural strength of the footing (per foot width) is

$$\phi M_n = \phi A_{s,\text{prov}} f_y \left(d - \frac{a}{2} \right)$$

$$= \frac{(0.9)(0.40 \text{ in}^2)\left(60 \frac{\text{kips}}{\text{in}^2}\right)\left(13.25 \text{ in} - \frac{0.588 \text{ in}}{2}\right)}{12 \frac{\text{in}}{\text{ft}}}$$

$$= \boxed{23.3 \text{ ft-kips/ft} > 5.12 \text{ ft-kips/ft} \quad [\text{OK}]}$$

Per ACI 318 Table 7.6.1.1, the minimum flexural reinforcement area, $A_{s,\text{min}}$, is

$$A_{s,\text{min}} = 0.0020 bh = (0.0020)(1 \text{ ft})\left(12 \frac{\text{in}}{\text{ft}}\right)(15 \text{ in})$$

$$= 0.36 < 0.40 \text{ in}^2 \quad [\text{OK}]$$

Check the minimum flexural steel area requirements of ACI 318 Sec. 9.6.1.2. The minimum area of tension is limited by

$$A_{s,\text{min}} = \max \begin{cases} \dfrac{3\sqrt{f_c'}\, b_w d}{f_y} = \dfrac{(3)\sqrt{4000 \frac{\text{lbf}}{\text{in}^2}} \times (1 \text{ ft})\left(12 \frac{\text{in}}{\text{ft}}\right)(13.25 \text{ in})}{\left(60 \frac{\text{kips}}{\text{in}^2}\right)\left(1000 \frac{\text{lbf}}{\text{kip}}\right)} \\ \qquad = 0.50 \text{ in}^2 \\[4pt] \dfrac{200 b_w d}{f_y} = \dfrac{(200)(1 \text{ ft})\left(12 \frac{\text{in}}{\text{ft}}\right)(13.25 \text{ in})}{\left(60 \frac{\text{kips}}{\text{in}^2}\right)\left(1000 \frac{\text{lbf}}{\text{kip}}\right)} \\ \qquad = 0.53 \text{ in}^2 \quad [\text{controls}] \end{cases}$$

Because the controlling value of 0.53 in^2 is greater than the provided flexural reinforcement area of 0.40 in^2, the requirements of ACI 318 Sec. 9.6.1.2 are not satisfied. However, from ACI 318 Sec. 9.6.1.3, the minimum steel area requirements of Sec. 9.6.1.2 do not need to be applied if $A_{s,\text{prov}}$ is at least one-third greater than the area required by analysis.

$$\frac{\phi M_n}{M_{u,\text{heel}}} = \frac{23.3 \frac{\text{ft-kips}}{\text{ft}}}{5.12 \frac{\text{ft-kips}}{\text{ft}}}$$

$$= 4.55 > 1.33$$

Therefore, $A_{s,\text{prov}}$ is at least one-third greater than the area required by analysis, and the minimum flexural reinforcement requirements are satisfied.

Check the maximum bar spacing, $s_{\max}$, per ACI 318 Sec. 24.3.2. Conservatively take the tensile stress in the reinforcement as

$$f_s = \frac{2}{3} f_y$$

$$= \left(\frac{2}{3}\right)\left(60 \frac{\text{kips}}{\text{in}^2}\right)\left(1000 \frac{\text{lbf}}{\text{kip}}\right)$$

$$= 40,000 \text{ lbf/in}^2$$

From ACI 318 Table 24.3.2, $s_{\max}$ is lesser of

$$s_{\max} = 15\left(\frac{40,000}{f_s}\right) - 2.5 c_c$$

$$= (15 \text{ in})\left(\frac{40,000 \frac{\text{lbf}}{\text{in}^2}}{40,000 \frac{\text{lbf}}{\text{in}^2}}\right) - (2.5)(1.5 \text{ in})$$

$$= 11.25 \text{ in} > 6 \text{ in} \quad [\text{OK}]$$

$$s_{\max} = 12\left(\frac{40,000}{f_s}\right) = 12 \text{ in} > 6 \text{ in} \quad [\text{ok}]$$

Therefore, no. 4 bars at 6 in o.c. are adequate for flexure in the heel.

(In this solution, the critical sections for shear and flexure were calculated at the least conservative sections permitted by ACI 318. Using other locations for the critical sections, such as at the centerline of the CMU stem, is also acceptable—as long as the locations of the assumed critical sections are conservative.)

43. (d) The concrete clear cover for the heel top bars was determined in part 43(c) to be $1\frac{1}{2}$ in. For the toe, the concrete will be cast against and permanently exposed to earth, so the cover for the bottom bars of the footing toe must be at least 3 in per ACI 318 Table 20.6.1.3.1.

From ACI 318 Sec. 13.2.8.1, the development of reinforcement in footings must follow the provisions of ACI 318 Chap. 25. From ACI 318 Sec. 25.4.1.1, the calculated tension in the flexural reinforcing bars must be developed beyond the critical section for flexure. In part 43(c), the critical section for flexure was determined to be halfway between the middle and edge of the CMU stem.

ACI 318 Eq. 25.4.2.3a gives the development length, l_d.

$$l_d = \left(\frac{3f_y\Psi_t\Psi_e\Psi_s}{40\lambda\sqrt{f_c'}\left(\dfrac{c_b + K_{tr}}{d_b}\right)}\right)d_b$$

From ACI 318 R25.4.2.3, c_b is the lesser of

$$c_b = \min\begin{cases} c_c + \dfrac{d_b}{2} = 1.5 \text{ in} + \dfrac{0.5 \text{ in}}{2} = 1.75 \text{ in} \quad \text{[controls]} \\ \dfrac{s}{2} = \dfrac{6 \text{ in}}{2} = 3 \text{ in} \end{cases}$$

From ACI 318 Sec. 25.4.2.3, $K_{tr} = 0$ for no transverse reinforcement.

From ACI 318 Sec. 25.4.2.3, the confinement term $(c_b + K_{tr})/d_b$ cannot be taken as greater than 2.5.

$$\frac{c_b + K_{tr}}{d_b} = \frac{1.75 \text{ in} + 0}{0.5 \text{ in}} = 3.5 > 2.5$$

Therefore, use $(c_b + K_{tr})/d_b = 2.5$.

At the heel, there are more than 12 in of fresh concrete cast below the bar. Therefore, from ACI 318 Sec. 25.4.2.4, $\Psi_{t,\text{heel}} = 1.3$.

Other variables are defined by ACI 318 Sec. 25.4.2.4 as follows: $\Psi_e = 1.0$ for uncoated reinforcement, $\Psi_s = 0.8$ for no. 4 bars, and $\lambda = 1.0$ for normal weight concrete.

The development length of the heel, $l_{d,\text{heel}}$, is

$$l_{d,\text{heel}} = \left(\frac{3f_y\Psi_{t,\text{heel}}\Psi_e\Psi_s}{40\lambda\sqrt{f_c'}\left(\dfrac{c_b + K_{tr}}{d_b}\right)}\right)d_b$$

$$= \left(\frac{(3)\left(60 \dfrac{\text{kips}}{\text{in}^2}\right)\left(1000 \dfrac{\text{lbf}}{\text{kip}}\right)(1.3)(1.0)(0.8)}{(40)(1.0)\sqrt{4000 \dfrac{\text{lbf}}{\text{in}^2}}\,(2.5)}\right)$$

$$\times(0.5 \text{ in})$$

$$= \boxed{15 \text{ in}}$$

For the bottom footing bars at the toe, $\Psi_t = 1.0$. Therefore, the development length of the toe, $l_{d,\text{toe}}$, is

$$l_{d,\text{toe}} = \frac{l_{d,\text{heel}}}{\Psi_{t,\text{heel}}} = \frac{15 \text{ in}}{1.3} = \boxed{12 \text{ in}}$$

(Also acceptable are other simplified and conservative expressions for calculating the development length (such as the expressions in ACI 318 Sec. 25.4.2.2 or ACI 318 Chap. 25 Commentary), as long as the assumptions are clearly stated. In other words, calculating a longer development length is not incorrect. The required development length could be reduced by the ratio $A_{s,\text{req}}/A_{s,\text{prov}}$ per ACI 318 Sec. 25.4.10.1, but cannot be less than 12 in per ACI 318 Sec. 25.4.2.1(b).)

Using a development length of 15 in, the total bar length for the heel bars, l_{top}, is

$$l_{\text{top}} = l_{\text{heel}} - c_c + \frac{t_{\text{wall}}}{4} + l_{d,\text{heel}}$$

$$= 4 \text{ ft} - \frac{1.5 \text{ in}}{12 \dfrac{\text{in}}{\text{ft}}} + \frac{9.625 \text{ in}}{(4)\left(12 \dfrac{\text{in}}{\text{ft}}\right)} + \frac{15 \text{ in}}{12 \dfrac{\text{in}}{\text{ft}}}$$

$$= \boxed{5.33 \text{ ft} \quad (5 \text{ ft } 4 \text{ in})}$$

Using a development length of 12 in, the total bar length for the toe bars, l_{bottom}, is

$$l_{\text{bottom}} = l_{\text{toe}} - c_c + \frac{t_{\text{wall}}}{4} + l_{d,\text{toe}}$$

$$= 2 \text{ ft} - \frac{1.5 \text{ in}}{12 \dfrac{\text{in}}{\text{ft}}} + \frac{9.625 \text{ in}}{(4)\left(12 \dfrac{\text{in}}{\text{ft}}\right)} + \frac{12 \text{ in}}{12 \dfrac{\text{in}}{\text{ft}}}$$

$$= \boxed{3.08 \text{ ft} \quad (3 \text{ ft } 1 \text{ in})}$$

Sketch a section through the reinforced footing that shows bar lengths and concrete clear cover.

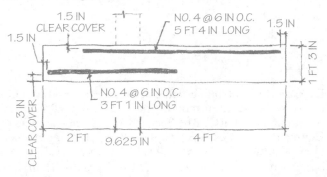

44. **(a)** From ASCE/SEI7 Eq. 7.3-1, the flat roof snow load, p_f, is

$$p_f = 0.7C_eC_tI_sp_g$$

From ASCE/SEI7 Table 7-2, for a partially exposed roof in terrain category C, the exposure factor, C_e, is

$$C_e = 1.0$$

From ASCE/SEI7 Table 7-3, for unheated structures, the thermal factor, C_t, is

$$C_t = 1.2$$

From ASCE/SEI7 Table 1.5-1, buildings that represent a low risk to human life in the event of failure are in risk category I. Therefore, from ASCE/SEI7 Table 1.5-2, the snow importance factor, I_s, is

$$I_s = 0.80$$

The flat roof snow load is

$$p_f = 0.7 C_e C_t I p_g = (0.7)(1.0)(1.2)(0.8)\left(60 \ \frac{\text{lbf}}{\text{ft}^2}\right)$$

$$= 40.32 \ \text{lbf/ft}^2$$

The roof slope, θ, is

$$\theta = \arctan \frac{l_{\text{opp}}}{l_{\text{adj}}} = \arctan \frac{8 \ \text{ft}}{16 \ \text{ft}} = 26.57°$$

The slope is more than 15°, so the roof is not considered "low-slope" per ASCE/SEI7 Sec. 7.3.4. Therefore, minimum load values for low-slope roofs may be ignored.

From ASCE/SEI7 Eq. 7.4-1, the sloped roof snow load, p_s, is given by

$$p_s = C_s p_f$$

Determine the roof slope factor, C_s. From ASCE/SEI7 Fig. 7-2c, for cold roofs with $C_t = 1.2$, an unobstructed, slippery surface, and a roof slope of $\theta = 26.57°$ (6 ft on 16 ft)

$$C_s = 0.79$$

The sloped roof snow load is

$$p_s = C_s p_f = (0.79)\left(40.32 \ \frac{\text{lbf}}{\text{ft}^2}\right)$$

$$= \boxed{31.85 \ \text{lbf/ft}^2 \quad (32 \ \text{lbf/ft}^2)}$$

44. (b) The sloped roof snow load acts on the horizontal projection of the roof, while the given dead load acts on the inclined roof area. Converting the roof dead load to the horizontal projection, and using the slope calculated in part 44(a), the total roof load $(D + S)$ is

$$q_{\text{tot}} = \frac{q_{\text{dead}}}{\cos\theta} + p_s = \frac{30 \ \dfrac{\text{lbf}}{\text{ft}^2}}{\cos 26.57°} + 31.85 \ \frac{\text{lbf}}{\text{ft}^2}$$

$$= 65.4 \ \text{lbf/ft}^2$$

For the interior truss joints (nodes 2, 3, and 4), the tributary area, $A_{t,\text{int}}$, is

$$A_{t,\text{int}} = s_{\text{truss}} s_{\text{purlin}} = (10 \ \text{ft})(8 \ \text{ft}) = 80 \ \text{ft}^2$$

For the end truss joints (nodes 1 and 5), the tributary area, $A_{t,\text{end}}$, is

$$A_{t,\text{end}} = s_{\text{truss}}\left(\frac{s_{\text{purlin}}}{2}\right) = (10 \ \text{ft})\left(\frac{8 \ \text{ft}}{2}\right) = 40 \ \text{ft}^2$$

The interior truss joint load, $P_{2,3,4}$, is

$$P_{2,3,4} = q_{\text{tot}} A_{t,\text{int}} = \frac{\left(65.4 \ \dfrac{\text{lbf}}{\text{ft}^2}\right)(80 \ \text{ft}^2)}{1000 \ \dfrac{\text{lbf}}{\text{kip}}} = \boxed{5.23 \ \text{kips}}$$

The exterior truss joint load, $P_{1,5}$, is

$$P_{1,5} = q_{\text{tot}} A_{t,\text{end}} = \frac{\left(65.4 \ \dfrac{\text{lbf}}{\text{ft}^2}\right)(40 \ \text{ft}^2)}{1000 \ \dfrac{\text{lbf}}{\text{kip}}} = \boxed{2.62 \ \text{kips}}$$

Due to symmetry, the vertical reactions, R, are both equal to half the total load on the truss.

$$R = \frac{\sum P_{1\text{-}5}}{2} = \frac{(2)(2.62 \ \text{kips}) + (3)(5.23 \ \text{kips})}{2}$$

$$= \boxed{10.5 \ \text{kips}}$$

The elevation of the truss is drawn as shown.

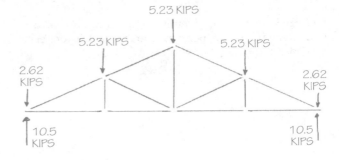

44. (c) Draw a free-body diagram of node 1.

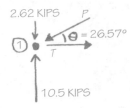

Using the value for θ calculated in part 44(a) and the values for R and $P_{1,5}$ calculated in part 44(b), the compression force, P, in member 1–2 is

$$P = \frac{R - P_{1,5}}{\sin \theta}$$

$$= \frac{(10.47 \text{ kips})\left(1000 \, \dfrac{\text{lbf}}{\text{kip}}\right) - (2.62 \text{ kips})\left(1000 \, \dfrac{\text{lbf}}{\text{kip}}\right)}{\sin 26.57°}$$

$$= 17{,}550 \text{ lbf}$$

Using the value for θ calculated in part 44(a), the tensile force, T, in member 1–6 is

$$T = P \cos \theta = (17{,}550 \text{ lbf}) \cos 26.57° = \boxed{15{,}700 \text{ lbf}}$$

Determine the smallest adequate size for member 1–6, assuming a nominal thickness of 3 in.

From NDS Sec. 3.8, the tensile stress, f_t, cannot exceed the adjusted tensile design value, F_t'.

$$F_t' \geq f_t$$

Using the factors from NDS Table 4.3.1, the adjusted tensile design value is

$$F_t' = F_t C_D C_M C_t C_F C_i$$

From the problem statement, $C_M = C_t = C_i = 1.0$.

From NDS Table 2.3.2, the load duration factor, C_D, for snow loads is

$$C_D = 1.15$$

The size factor, C_F, is given in NDS Supp. Table 4A. For a no. 1 sawn lumber member, the size factor for tension depends on the member width.

Try a 3×8 member, which has a size factor for tension of $C_F = 1.2$.

$$F_t' = F_t C_D C_M C_t C_F C_i$$

$$= \left(675 \, \frac{\text{lbf}}{\text{in}^2}\right)(1.15)(1.0)(1.0)(1.2)(1.0)$$

$$= 932 \text{ lbf/in}^2$$

The cross-sectional area, A, of a 3×8 member is given in NDS Supp. Table 1B as 18.13 in².

The calculated tensile stress for a 3×8 member is

$$f_t = \frac{T}{A} = \frac{15{,}700 \text{ lbf}}{18.13 \text{ in}^2}$$

$$= \boxed{866 \text{ lbf/in}^2 \leq 932 \text{ lbf/in}^2 \quad \text{[OK]}}$$

A 3×8 member has a demand-to-capacity ratio very close to 1.0.

$$\frac{\text{demand}}{\text{capacity}} = \frac{866 \, \dfrac{\text{lbf}}{\text{in}^2}}{932 \, \dfrac{\text{lbf}}{\text{in}^2}}$$

$$= 0.93$$

Therefore, by inspection, a 3×6 member will not be adequate.

> A 3×8 member is the smallest adequate size for member 1–6.

44. (d) From illustration III, only member 1–2 bears on the perimeter wall. So, the full truss reaction must be considered, and the bearing stress, f_θ, on the double 2×8 member cannot exceed the adjusted bearing design value, F_θ'.

$$F_\theta' = \frac{F_c^* F_{c\perp}'}{F_c^* \sin^2 \theta + F_{c\perp}' \cos^2 \theta}$$

NDS Table 4.3.1 lists applicable adjustment factors.

F_c^*, the reference compression design value parallel to grain multiplied by all applicable adjustment factors except C_P, is

$$F_c^* = F_c C_D C_M C_t C_F C_i$$

$F'_{c\perp}$, the reference compression design value perpendicular to grain multiplied by all applicable adjustment factors, is

$$F'_{c\perp} = F_c C_M C_t C_i C_b$$

From the problem statement, $C_M = C_t = C_i = 1.0$.

From NDS Table 2.3.2, the load duration factor, C_D, for snow loads is

$$C_D = 1.15$$

For an 8 in deep member, NDS Supp. Table 4A gives a size factor for compression of

$$C_F = 1.05$$

From NDS Sec. 3.10.4, the bearing area factor, C_b, is only applicable for bearing away from the end of a member. So,

$$C_b = 1.0$$

Calculate the design values.

$$\begin{aligned}
F_c^* &= F_c C_D C_M C_t C_F C_i \\
&= \left(1500\ \frac{\text{lbf}}{\text{in}^2}\right)(1.15)(1.0)(1.0)(1.05)(1.0) \\
&= 1811\ \text{lbf/in}^2
\end{aligned}$$

$$\begin{aligned}
F'_{c\perp} &= F_{c\perp} C_M C_t C_i C_b \\
&= \left(625\ \frac{\text{lbf}}{\text{in}^2}\right)(1.0)(1.0)(1.0)(1.0) \\
&= 625\ \text{lbf/in}^2
\end{aligned}$$

From NDS Sec. 3.10.3, θ is the angle between the direction of the load and the direction of the grain. In this case, the load is the vertical reaction of the truss, and the grain runs along the length of the member. Therefore, the angle is

$$\begin{aligned}
\theta &= \arctan \frac{l_{\text{opp}}}{l_{\text{adj}}} \\
&= \arctan \frac{16\ \text{ft}}{8\ \text{ft}} \\
&= 63.4°
\end{aligned}$$

The adjusted bearing design value is

$$\begin{aligned}
F'_\theta &= \frac{F_c^* F'_{c\perp}}{F_c^* \sin^2\theta + F'_{c\perp}\cos^2\theta} \\
&= \frac{\left(1811\ \dfrac{\text{lbf}}{\text{in}^2}\right)\left(625\ \dfrac{\text{lbf}}{\text{in}^2}\right)}{\left(1811\ \dfrac{\text{lbf}}{\text{in}^2}\right)(\sin 63.4)^2 + \left(625\ \dfrac{\text{lbf}}{\text{in}^2}\right)(\cos 63.4)^2} \\
&= 719\ \text{lbf/in}^2
\end{aligned}$$

The calculated bearing stress is the truss reaction divided by the bearing area of member 1–2. The truss reaction is given in the problem statement as 15 kips.

$$f_\theta = \frac{R}{2bl_b}$$

The width, b, of a single 2×8 member is 1.5 in. Equating F'_θ and f_θ, the required bearing length, l_b, is

$$l_b = \frac{R}{F'_\theta 2b} = \frac{(15\ \text{kips})\left(1000\ \dfrac{\text{lbf}}{\text{kip}}\right)}{\left(719\ \dfrac{\text{lbf}}{\text{in}^2}\right)(2)(1.5\ \text{in})} = \boxed{7.0\ \text{in}}$$

44. **(e)** From NDS Sec. 10.2.2, the total adjusted design value, Z', for a connection with multiple fasteners is the sum of the adjusted design values for each individual fastener. Using NDS Table 10.3.1, the adjusted design value for a dowel-type fastener is

$$Z' = Z C_D C_M C_t C_g C_\Delta C_{\text{eg}} C_{\text{di}} C_{\text{tn}}$$

For this connection, the end grain factor, C_{eg}, diaphragm factor, C_{di}, and toe-nail factor, C_{tn}, do not apply. Therefore, $C_{\text{eg}} = C_{\text{di}} = C_{\text{tn}} = 1.0$. Since all other adjustment factors except for C_Δ also equal 1.0, the adjusted lateral design value for a single bolt is

$$Z' = Z C_\Delta$$

The reference design value, Z, can be obtained from NDS Table 11A. For the given connection configuration, $t_m = t_s = 3.5$ in. The bolt bearing is perpendicular to the grain of side member 1–6. So, for a Douglas fir-larch (SG $= 0.50$) and a bolt diameter of $D = \frac{1}{2}$ in, the bearing strength for loading perpendicular to the side member is

$$Z = Z_s = 490\ \text{lbf}$$

From NDS Sec. 11.5.1, the geometry factor, C_Δ, is the smallest value considering edge distance, end distance, and spacing requirements. The smallest geometry factor for any fastener in a group must be applied to all fasteners in the group.

Check spacing requirements. From NDS Table 11.5.1B, the minimum spacing for fasteners in a row for $C_\Delta = 1.0$ is

$$4D = (4)(0.5 \text{ in})$$
$$= 2 \text{ in} \le 3.25 \text{ in} \quad [\text{OK}]$$

From NDS Table 11.5.1D, for loading perpendicular to grain and for $l/D = 3.5 \text{ in}/0.5 \text{ in} = 7 \ge 6$, the minimum spacing between rows is

$$5D = (5)(0.5 \text{ in})$$
$$= 2.5 \text{ in} \le 3.25 \text{ in} \quad [\text{OK}]$$

The bolts bear perpendicular to the grain for the side member, and they bear parallel to the grain for the main member. Therefore, both edge and end distance requirements must be checked.

From NDS Table 11.5.1C, for bearing perpendicular to grain at a loaded edge, the minimum edge distance requirement is

$$4D = (4)(0.5 \text{ in})$$
$$= 2 \text{ in} \le 2 \text{ in} \quad [\text{OK}]$$

The bolts are loaded parallel to the grain of the main member with the fasteners bearing toward the member end. From NDS Table 11.5.1A, for softwoods with these conditions, the minimum end distance for $C_\Delta = 1.0$ is

$$7D = (7)(0.5 \text{ in})$$
$$= 3.5 \text{ in}$$

3.5 in is greater than the provided end distance of 2 in for the two bottom fasteners. Therefore, the geometry factor is

$$C_\Delta = \frac{\text{actual end distance}}{\text{minimum end distance for } C_\Delta = 1.0}$$
$$= \frac{2 \text{ in}}{3.5 \text{ in}}$$
$$= 0.571$$

For 4 bolts, the adjusted design value for the connection is

$$Z' = nZC_\Delta = (4)(490 \text{ lbf})(0.571)$$
$$= \boxed{1120 \text{ lbf}}$$

Solutions
Lateral Forces Component: Breadth Module Exam

45. Parapet wind loads for components and cladding on an enclosed building with a roof height less than 160 ft are calculated using ASCE/SEI7 Sec. 30.7.1.2. Pressures are determined from ASCE/SEI7 Table 30.7-2 with adjustments per ASCE/SEI7 Eq. 30.7-1.

$$p = p_{table}(EAF)(RF)K_{zt}$$

The exposure adjustment factor, EAF, for exposure category C is 1.0.

The effective area reduction factor, RF, for an effective wind area of 10 ft² is 1.0.

The topographical factor, K_{zt}, is given as 1.0.

Therefore, $p = p_{table}$, and the parapet wind pressures can be read directly from ASCE/SEI7 Table 30.7-2.

From ASCE/SEI7 Sec. 30.7.1.2, the height, h, is to the top of the parapet.

$$h = \sum z_{floors} = 14 \text{ ft} + (3)(11 \text{ ft}) + 3 \text{ ft}$$
$$= 50 \text{ ft}$$

The application of the wind loads is illustrated in ASCE/SEI7 Fig. 30.7-1.

For the windward parapet (load case A), the windward parapet pressure, p_1, is determined using the positive wall pressure at zone 5 for a corner zone. From ASCE/SEI7 Table 30.7-2 for v = 130 mph and h = 50 ft with a flat roof, the positive wall pressure (load case 2) is

$$p_1 = 43.4 \text{ lbf/ft}^2$$

($p_1 = -79.6 \text{ lbf/ft}^2$ corresponds to the negative wall pressure, so it is not applicable for the windward parapet.)

The leeward pressure on the windward parapet, p_2, is determined using the negative roof pressure at zone 3 for a corner zone. From ASCE/SEI7 Table 30.7-2 for v = 130 mph and h = 50 ft with a flat roof,

$$p_2 = -136 \text{ lbf/ft}^2$$

From ASCE/SEI7 Sec. 30.7.1.2, the total pressure on the windward parapet is the sum of the absolute values of the windward and leeward pressures.

$$p = |p_1| + |p_2|$$
$$= \left| 43.4 \frac{\text{lbf}}{\text{ft}^2} \right| + \left| -136 \frac{\text{lbf}}{\text{ft}^2} \right|$$
$$= 179.4 \text{ lbf/ft}^2 \quad (180 \text{ lbf/ft}^2)$$

The answer is (C).

46. For strength design, use the following IBC Sec. 1605.2 load combinations to find the maximum factored axial load, P_u. (ASCE/SEI7 Sec. 2.3.2 could also be used.)

From IBC Eq. 16-1,

$$P_u = 1.4D = (1.4)(200 \text{ kips})$$
$$= 280 \text{ kips} \quad \text{[does not control]}$$

From IBC Eq. 16-2,

$$P_u = 1.2D + 1.6L = (1.2)(200 \text{ kips}) + (1.6)(75 \text{ kips})$$
$$= 360 \text{ kips} \quad \text{[does not control]}$$

Ignoring snow loads, from IBC Eq. 16-5,

$$P_u = 1.2D + 1.0E + f_1L$$

IBC Sec. 1605.2 gives f_1 as 1.0 for parking garages. E accounts for the horizontal and vertical seismic load effects, as described in ASCE/SEI7 Sec. 12.4.2.

$$E = E_h + E_v = \rho Q_E + 0.2S_{DS}D$$
$$= (1.3)(25 \text{ kips}) + (0.2)(1.9)(200 \text{ kips})$$
$$= 109 \text{ kips}$$

Therefore,

$$P_u = 1.2D + 1.0E + f_1L$$
$$= (1.2)(200 \text{ kips}) + (1.0)(109 \text{ kips}) + (1.0)(75 \text{ kips})$$
$$= 424 \text{ kips} \quad (420 \text{ kips}) \quad \text{[controls]}$$

Alternatively, for IBC Eq. 16-5, the loads can be input directly into ASCE/SEI7 Sec. 12.4.2.3 load combination 5. Ignoring snow loads,

$$P_u = (1.2 + 0.2S_{DS})D + \rho Q_E + L$$
$$= (1.2 + (0.2)(1.9))(200 \text{ kips}) + (1.3)(25 \text{ kips})$$
$$\quad + 75 \text{ kips}$$
$$= 424 \text{ kips} \quad (420 \text{ kips}) \quad [\text{controls}]$$

The answer is (D).

47. The calculation of seismic active earth pressure is covered in AASHTO Sec. 11.6.5.3. The seismic active pressure coefficient, K_{AE}, can be determined by using one of the methods outlined in AASHTO A11.3. (See AASHTO Fig. A11.3.2-2 for a chart that can be used to determine K_{AE} using the Mononobe-Okabe method.) However, for this question, K_{AE} is given in the problem statement. Therefore, the total horizontal seismic active force, P_{AE}, can be determined directly from AASHTO Eq. 11.6.5.3-2 as

$$P_{AE} = 0.5\gamma h^2 K_{AE} = (0.5)\left(0.130 \ \frac{\text{kips}}{\text{ft}^3}\right)(15 \text{ ft})^2(0.37)$$
$$= 5.4 \text{ kips/ft}$$

The answer is (D).

48. The deflected shape of a moment frame will maintain right angles between the columns and the beams. For a moment frame pinned at the base, the columns will be in single curvature, and the beam will be in double curvature.

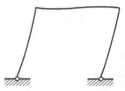

The answer is (A).

49. A site's seismic design category is determined using ASCE/SEI7 Sec. 11.6. However, first the site classification must be found. From ASCE/SEI7 Sec. 20.4.1, the average shear wave velocity, $\bar{v}_s$, is

$$\bar{v}_s = \frac{\sum_{i=1}^{n} d_i}{\sum_{i=1}^{n} \dfrac{d_i}{v_{si}}} = \frac{100 \text{ ft}}{\dfrac{20 \text{ ft}}{900 \ \dfrac{\text{ft}}{\text{sec}}} + \dfrac{30 \text{ ft}}{1000 \ \dfrac{\text{ft}}{\text{sec}}} + \dfrac{50 \text{ ft}}{1300 \ \dfrac{\text{ft}}{\text{sec}}}}$$
$$= 1103 \text{ ft/sec}$$

Using ASCE/SEI7 Table 20.3.1, for values of $\bar{v}_s$ between 600 ft/sec and 1200 ft/sec, the soil is classified as stiff soil, site class D.

Then, from ASCE/SEI7 Table 11.4-1, for site class D and using straight-line interpolation between $S_S \leq 0.25$ and $S_S = 0.5$, the short-period site coefficient is $F_a = 1.56$.

From ASCE/SEI7 Table 11.4-2, for site class D and $S_1 \leq 0.1$, the long-period site coefficient is $F_v = 2.4$.

Adjusting for site class effects, find the maximum considered earthquake spectral response acceleration for short periods, S_{MS}, and at 1 sec, S_{M1}. From ASCE/SEI7 Eq. 11.4-1 and Eq. 11.4-2,

$$S_{MS} = F_a S_S = (1.56)(0.30) = 0.468$$
$$S_{M1} = F_v S_1 = (2.4)(0.080) = 0.192$$

Find the design spectral response acceleration parameters, S_{DS} and S_{D1}. From ASCE/SEI7 Eq. 11.4-3 and Eq. 11.4-4,

$$S_{DS} = \tfrac{2}{3}S_{MS} = \left(\frac{2}{3}\right)(0.468) = 0.312$$
$$S_{D1} = \tfrac{2}{3}S_{M1} = \left(\frac{2}{3}\right)(0.192) = 0.128$$

In ASCE/SEI7 Table 1.5-1, essential facilities are classified as risk category IV. Since $0.167 \leq S_{DS} \leq 0.33$, and $0.067 \leq S_{D1} \leq 0.133$, from ASCE/SEI7 Table 11.6-1 and Table 11.6-2, the police station is classified as seismic design category C.

The answer is (C).

50. The seismic base shear is determined per ASCE/SEI7 Sec. 12.8. The fundamental period of the building, T_a, can be approximated using ASCE/SEI7 Eq. 12.8-7. From ASCE/SEI7 Table 12.8-2, $C_t = 0.028$ and $x = 0.8$ for steel moment-resisting frames.

$$T_a = C_t h_n^x = (0.028)(300 \text{ ft})^{0.8} = 2.68 \text{ sec}$$

(ASCE/SEI7 Eq. 12.8-8 cannot be used to determine the approximate fundamental period because the structure exceeds 12 stories in height.)

From ASCE/SEI7 Table 12.2-1, case C.1, a building with steel special moment frames has a response modification coefficient, R, of 8.

Per ASCE/SEI7 Table 1.5-1, an office building is in risk category II. ASCE/SEI7 Table 1.5-2 gives a seismic importance factor, I_e, of 1.00.

Use ASCE/SEI7 Eq. 12.8-2 to find the seismic response coefficient, C_s.

$$C_s = \frac{S_{DS}}{\dfrac{R}{I_e}} = \frac{1.0}{\dfrac{8}{1.0}}$$

$$= 0.125 \quad \text{[does not control]}$$

When the fundamental period is less than the long-period transition period $(T < T_L)$, C_s cannot exceed the value calculated using ASCE/SEI7 Eq. 12.8-3.

$$C_s = \frac{S_{D1}}{T\left(\dfrac{R}{I_e}\right)} = \frac{0.50}{(2.68 \text{ sec})\left(\dfrac{8}{1.0}\right)}$$

$$= 0.0233 \quad \text{[does not control]}$$

Using ASCE/SEI7 Eq. 12.8-5, C_s cannot be taken as less than.

$$C_s = 0.044 S_{DS} I_e \ge 0.01 = (0.044)(1.0)(1.0)$$
$$= 0.044 \quad \text{[controls]}$$

The mapped spectral response coefficient at 1 sec, S_1, is less than 0.60; therefore, ASCE/SEI7 Eq. 12.8-6 does not apply.

From ASCE/SEI7 Eq. 12.8-1, the seismic base shear, V, is

$$V = C_s W = (0.044)(7500 \text{ kips}) = 330 \text{ kips}$$

The answer is (B).

51. ASCE/SEI7 Sec. 12.8.3 describes vertical distribution of seismic forces. Per ASCE/SEI7 Eq. 12.8-11, the seismic force at the third floor, F_3, is

$$F_3 = C_{v3} V$$

The seismic response coefficient, C_s, is given as 0.048. W is the total seismic weight. From ASCE/SEI7 Eq. 12.8-1, the seismic base shear, V, of the structure is

$$V = C_s W$$
$$= (0.048)(350 \text{ kips} + 400 \text{ kips} + 300 \text{ kips})$$
$$= 50.4 \text{ kips}$$

The vertical distribution factor, C_{vx}, for the third floor is given by ASCE/SEI7 Eq. 12.8-12. For structures having

a fundamental period of 0.5 sec or less, k is 1.0. w_i is the seismic weight at story i, and h_i is the height of story i.

$$C_{v3} = \frac{w_3 h_3^{k}}{\sum_{i=1}^{n} w_i h_i^{k}}$$

$$= \frac{(300 \text{ kips})(15 \text{ ft} + 12 \text{ ft} + 12 \text{ ft})^{1.0}}{(350 \text{ kips})(15 \text{ ft})^{1.0} + (400 \text{ kips})(15 \text{ ft} + 12 \text{ ft})^{1.0}}$$
$$\qquad \overline{+(300 \text{ kips})(15 \text{ ft} + 12 \text{ ft} + 12 \text{ ft})^{1.0}}$$

$$= 0.422$$

The seismic lateral force at the third floor, F_3, is

$$F_3 = C_{v3} V$$
$$= (0.422)(50.4 \text{ kips})$$
$$= 21.3 \text{ kips} \quad (21 \text{ kips})$$

The answer is (C).

52. Seismic design requirements for nonbuilding structures are covered in ASCE/SEI7 Chap. 15. ASCE/SEI7 Sec. 15.4.2 states that a nonbuilding structure with a fundamental period of less than 0.06 sec can be considered rigid. The lateral seismic base shear, V, for rigid nonbuilding structures is calculated using ASCE/SEI7 Eq. 15.4-5.

$$V = 0.30 S_{DS} W I_e = (0.30)(0.30)(20 \text{ kips})(1.0)$$
$$= 1.8 \text{ kips}$$

The seismic shear force acts at the centroid of the billboard. Therefore, the moment, M, at the base of the pipe is equal to the base shear multiplied by the height, h, of the billboard's center.

$$M = Vh = (1.8 \text{ kips})(8 \text{ ft}) = 14.4 \text{ ft-kips}$$

From the AISC *Steel Construction Manual* Table 1-14, the section modulus, S, of an 8 in standard pipe is 15.8 in^3.

The maximum flexural stress, f_b, due to the lateral seismic force is calculated as the moment at the base of the pipe divided by the section modulus.

$$f_b = \frac{M}{S} = \frac{(14.4 \text{ ft-kips})\left(12 \dfrac{\text{in}}{\text{ft}}\right)}{15.8 \text{ in}^3} = 11 \text{ kips/in}^2$$

The answer is (C).

53. From ASCE/SEI7 Sec. 11.6, buildings in risk category I, II, or III with $S_1 \ge 0.75$ are in seismic design category E.

ASCE/SEI7 Table 12.2-1 gives building height limits for various seismic force-resisting systems. For seismic design category E, special reinforced concrete shear walls (ASCE/SEI7 Table 12.2-1, case A.1 or case B.4) and steel eccentrically braced frames (ASCE/SEI7 Table 12.2-1, case B.1) have a building height limit of 160 ft. This height limit may be increased to 240 ft, as the building meets the requirements of ASCE/SEI7 Sec. 12.2.5.4. However, the 240 ft height limit would not be sufficient for a 40-story office building.

For special reinforced concrete moment frames (ASCE/SEI7 Table 12.2-1, case C.5), the building height is not limited. Therefore, the only listed system that could be used for the building is a special reinforced concrete moment frame.

The answer is (A).

54. Calculate the bridge lateral stiffness, K, per AASHTO Eq. C4.7.4.3.2c-1.

$$K = \frac{p_o L}{v_{s,\max}} = \frac{\left(1.0 \; \dfrac{\text{kip}}{\text{ft}}\right)(120 \text{ ft})}{0.0040 \text{ ft}} = 30{,}000 \text{ kips/ft}$$

From AASHTO Eq. C4.7.4.3.2c-2, the bridge weight, W, is

$$W = \int w(x)\,dx = wL = \left(7.5 \; \frac{\text{kips}}{\text{ft}}\right)(120 \text{ ft})$$
$$= 900 \text{ kips}$$

AASHTO Eq. C4.7.4.3.2c-3 gives the period of the bridge, T_m, as

$$T_m = 2\pi \sqrt{\frac{W}{gK}} = 2\pi \sqrt{\frac{900 \text{ kips}}{\left(32.2 \; \dfrac{\text{ft}}{\text{sec}^2}\right)\left(30{,}000 \; \dfrac{\text{kips}}{\text{ft}}\right)}}$$
$$= 0.192 \text{ sec}$$

From the given design response spectrum, the period of the bridge is greater than T_0 and less than T_S. Therefore, AASHTO Eq. 3.10.4.2-4 gives the seismic response coefficient, C_{sm}, as

$$C_{sm} = S_{DS} = 0.80$$

From AASHTO Eq. C4.7.4.3.2c-4, the equivalent uniform static transverse seismic load, p_e, is

$$p_e = \frac{C_{sm}W}{L} = \frac{(0.80)(900 \text{ kips})}{120 \text{ ft}} = 6.0 \text{ kips/ft}$$

The answer is (B).

55. Because the superstructure is restrained laterally only by the column, the column must resist the entire lateral seismic load. The lateral earthquake load, F, is the same in both the transverse and longitudinal directions. It is equal to the elastic seismic response coefficient, C_{sm}, multiplied by the total weight of the superstructure, wL.

$$F_x = F_y = C_{sm}wL = (0.50)\left(5.25 \; \frac{\text{kips}}{\text{ft}}\right)(80 \text{ ft})$$
$$= 210 \text{ kips}$$

For a column fixed at the top and at the base, the elastic moment, M_{ex}, in the column is

$$M_{ex} = M_{ey} = F\left(\frac{H}{2}\right) = (210 \text{ kips})\left(\frac{18 \text{ ft}}{2}\right)$$
$$= 1890 \text{ ft-kips}$$

AASHTO Sec. 3.10.8 states that the force effects in two orthogonal directions must be combined to determine the total load effect. Considering 100% of the moment in one direction plus 30% of the moment in the perpendicular direction, the resultant elastic moment for a circular column is

$$M_e = \sqrt{\left(1.0 M_{ex}\right)^2 + \left(0.30 M_{ey}\right)^2}$$
$$= \sqrt{\left((1.0)(1890 \text{ ft-kips})\right)^2 + \left((0.30)(1890 \text{ ft-kips})\right)^2}$$
$$= 1973 \text{ ft-kips}$$

Per AASHTO Sec. 3.10.7, seismic design force effects are determined by dividing the force effects resulting from elastic analysis by the appropriate response modification factor, R. From AASHTO Table 3.10.7.1-1, for a single column of an essential bridge, the response modification factor is 2.0. Therefore, the design moment, M_u, for the earthquake loading is

$$M_u = \frac{M_e}{R} = \frac{1973 \text{ ft-kips}}{2.0} = 987 \text{ ft-kips} \quad (990 \text{ ft-kips})$$

The answer is (B).

56. From the problem design criteria, the relative rigidities, R, of the walls are proportional to their lengths. (This is generally valid for shear walls with the same thickness and when the height to length ratio is less than 0.30.) So,

$$R_A = 50$$
$$R_B = 60$$

Taking the origin at the west end of the building, shear wall A is located at $x_A = 0$ ft and shear wall B is located at $x_B = 150$ ft. Ignoring out-of-plane stiffness, shear wall C does not contribute to the stiffness in the east-west direction. Therefore, the x coordinate of the center of rigidity, x_{CR}, is at

$$x_{CR} = \frac{R_A x_A + R_B x_B}{R_A + R_B} = \frac{(50)(0 \text{ ft}) + (60)(150 \text{ ft})}{50 + 60}$$
$$= 81.8 \text{ ft}$$

Since shear wall C is the only element resisting loads in the east-west direction, the y-coordinate of the center of rigidity is aligned with shear wall C. Therefore, shear wall C does not contribute to the torsional stiffness.

The polar moment of inertia, J, of the shear walls is

$$J = R_A(x_A - x_{CR})^2 + R_B(x_B - x_{CR})^2$$
$$= (50)(0 \text{ ft} - 81.8 \text{ ft})^2 + (60)(150 \text{ ft} - 81.8 \text{ ft})^2$$
$$= 613{,}600 \text{ ft}^2$$

The total torsional moment, M_T, is the sum of the moments due to 1) the eccentricity between the center of rigidity and the center of mass and 2) a 5% accidental torsion per ASCE/SEI7 Sec. 12.8.4.2.

$$M_T = V(x_{CR} - x_{CM}) + V(0.05)L$$
$$= (40 \text{ kips})(81.8 \text{ ft} - 75 \text{ ft})$$
$$\quad + (40 \text{ kips})(0.05)(150 \text{ ft})$$
$$= 572 \text{ ft-kips}$$

The shear force in wall A, V_A, is

$$V_A = \frac{VR_A}{R_A + R_B} + \frac{M_T R_A(x_{CR} - x_A)}{J}$$
$$= \frac{(40 \text{ kips})(50)}{50 + 60} + \frac{(572 \text{ ft-kips})(50)(81.8 \text{ ft} - 0 \text{ ft})}{613{,}600 \text{ ft}^2}$$
$$= 22 \text{ kips}$$

The answer is (D).

57. Use the process of elimination to determine which structural irregularity exists for the building.

ASCE/SEI7 Sec. 12.3.2.2 covers vertical structural irregularities.

From ASCE/SEI7 Table 12.3-2, a vertical geometric irregularity exists where the horizontal dimension of the seismic force-resisting system in any story is more than 130% of that in an adjacent story. The horizontal dimension of the building's first and second stories is

60 ft, and the horizontal dimension of the third and fourth stories is 50 ft, so

$$\frac{\text{width of second story}}{\text{width of third story}} = \frac{60 \text{ ft}}{50 \text{ ft}} = 1.20$$
$$= 120\% < 130\% \quad [\text{OK}]$$

A vertical geometric irregularity does not exist for the building.

From ASCE/SEI7 Table 12.3-2, a stiffness-soft story irregularity exists when a story's lateral stiffness is less than 70% of that in the story above it, or less than 80% of the average stiffness of the three stories above it. A stiffness-extreme soft story irregularity exists when a story's lateral stiffness is less than 60% of that in the story above it, or less than 70% of the average stiffness of the three stories above it.

Calculate each story's lateral stiffness to determine if either irregularity exists for the building.

From the problem statement, each story's lateral stiffness, R, is inversely proportional to its drift ratio, δ_{xe}/h_{sx}.

$$R = \frac{1}{\text{drift ratio}} = \frac{1}{\dfrac{\delta_{xe}}{h_{sx}}} = \frac{h_{sx}}{\delta_{xe}}$$

Find the relative lateral stiffness of each story.

$$R_1 = \frac{h_{s1}}{\delta_{1e}} = \frac{(15 \text{ ft})\left(12 \ \frac{\text{in}}{\text{ft}}\right)}{0.91 \text{ in}} = 198$$

$$R_2 = \frac{h_{s2}}{\delta_{2e} - \delta_{1e}} = \frac{(10 \text{ ft})\left(12 \ \frac{\text{in}}{\text{ft}}\right)}{1.30 \text{ in} - 0.91 \text{ in}} = 308$$

$$R_3 = \frac{h_{s3}}{\delta_{3e} - \delta_{2e}} = \frac{(10 \text{ ft})\left(12 \ \frac{\text{in}}{\text{ft}}\right)}{1.78 \text{ in} - 1.30 \text{ in}} = 250$$

$$R_4 = \frac{h_{s4}}{\delta_{4e} - \delta_{3e}} = \frac{(10 \text{ ft})\left(12 \ \frac{\text{in}}{\text{ft}}\right)}{2.23 \text{ in} - 1.78 \text{ in}} = 267$$

Find the average lateral stiffness of stories two to four.

$$R_{2-4} = \frac{R_2 + R_3 + R_4}{3} = \frac{308 + 250 + 267}{3} = 275$$

The first story is less stiff than the stories above it, so either a soft story or extreme soft story irregularity is possible.

The largest difference in lateral stiffness will be between the first and second stories. Calculate the ratio of the first story's lateral stiffness to the second story's lateral stiffness.

$$\frac{R_1}{R_2} = \frac{198}{308} \times 100\% = 64\%$$

This is less than 70%, but greater than 60%, so a soft story irregularity, but not an extreme soft story irregularity, exists at the first story.

Comparing the stiffness of the first story to the stiffness of the average of the three stories above it gives

$$\frac{R_1}{R_{2-4}} = \frac{198}{275} \times 100\% = 72\%$$

This is less than 80%, but greater than 70%. So, a soft story irregularity, but not an extreme soft story irregularity, still exists.

ASCE/SEI7 Sec. 12.3.2.2, exception 1 does not apply for

$$\frac{R_2}{R_1} = \frac{308}{198} \times 100\% = 156\% > 130\%$$

Therefore, only a stiffness-soft story irregularity exists.

The answer is (B).

58. The building wall and parapet can be analyzed as a beam with an overhang. The wind load on the roof diaphragm is the reaction of the uniformly loaded "beam" at the roof level.

Using the shear and moment diagrams from AISC *Steel Construction Manual* Table 3-23, case 24, calculate the reaction at the roof level, R. w is the wind load, h is the roof height, and a is the parapet height.

$$R = \left(\frac{w}{2h}\right)(h + a)^2 = \left(\frac{30 \ \frac{\text{lbf}}{\text{ft}^2}}{(2)(20 \ \text{ft})}\right)(20 \ \text{ft} + 3 \ \text{ft})^2$$

$$= 397 \ \text{lbf/ft}$$

The diaphragm is simply supported and spans between the east and west walls. The diaphragm moment, M, at

point C (at $x = 20$ ft from the east wall) is given in AISC *Steel Construction Manual* Table 3-23, case 1, as

$$M = \left(\frac{Rx}{2}\right)(l - x) = \left(\frac{\left(397 \ \frac{\text{lbf}}{\text{ft}}\right)(20 \ \text{ft})}{(2)\left(1000 \ \frac{\text{lbf}}{\text{kip}}\right)}\right)(50 \ \text{ft} - 20 \ \text{ft})$$

$$= 119 \ \text{ft-kips}$$

The unfactored chord force, T, at point C is the diaphragm moment divided by the building width, B.

$$T = \frac{M}{B} = \frac{119 \ \text{ft-kips}}{15 \ \text{ft}} = 7.9 \ \text{kips}$$

The answer is (B).

Alternative Solution

This problem can also be solved by approximating the diaphragm force. Use a tributary width of half the building height plus the parapet height.

The diaphragm reaction, R, at the roof is

$$R = w\left(\frac{h}{2} + a\right) = \left(30 \ \frac{\text{lbf}}{\text{ft}^2}\right)\left(\frac{20 \ \text{ft}}{2} + 3 \ \text{ft}\right) = 390 \ \text{lbf/ft}$$

The diaphragm moment, M, is

$$M = \left(\frac{Rx}{2}\right)(l - x) = \left(\frac{\left(390 \ \frac{\text{lbf}}{\text{ft}}\right)(20 \ \text{ft})}{(2)\left(1000 \ \frac{\text{lbf}}{\text{kip}}\right)}\right)(50 \ \text{ft} - 20 \ \text{ft})$$

$$= 117 \ \text{ft-kips}$$

The resulting chord force, T, at point C is

$$T = \frac{M}{B} = \frac{117 \ \text{ft-kips}}{15 \ \text{ft}} = 7.8 \ \text{kips} \quad (7.9 \ \text{kips})$$

The answer is (B).

59. The net design wind pressure for an open building with a pitched roof is given by ASCE/SEI7 Eq. 27.4-3 as

$$p = q_h G C_N$$

The mean roof height is

$$h = 8 \ \text{ft} + \frac{5 \ \text{ft}}{2} = 10.5 \ \text{ft}$$

From ASCE/SEI7 Table 27.3-1, the velocity pressure exposure coefficient, K_z, for exposure category C and at height above the ground of less than 15 ft, is 0.85.

From ASCE/SEI7 Table 26.6-1, the directionality factor, K_d, for the main wind force resisting system of buildings is 0.85.

Therefore, use ASCE/SEI7 Eq. 27.3-1 to find the velocity pressure. (ASCE/SEI7 uses V for velocity, whereas this solution uses v.)

$$q_h = q_z = 0.00256 K_z K_{zt} K_d v^2$$
$$= (0.00256)(0.85)(1.0)(0.85)\left(170 \ \frac{mi}{hr}\right)^2$$
$$= 53.5 \ lbf/ft^2$$

The net pressure coefficient, C_N, is determined from ASCE/SEI7 Fig. 27.4-5 using the roof angle, θ.

The roof angle, θ, is

$$\theta = \arctan \frac{5 \ ft}{12 \ ft} = 22.6° \quad [\text{use } 22.5°]$$

For clear wind flow, the worst case net pressure coefficient for the windward half of the roof surface is

$$C_{NW} = 1.1$$

The net wind pressure on the windward half of the roof is

$$p_{NW} = q_h G C_{NW} = \left(53.5 \ \frac{lbf}{ft^2}\right)(0.85)(1.1)$$
$$= 50 \ lbf/ft^2$$

The answer is (A).

60. The base wind pressure on superstructure girders, $P_{B,super}$, for evaluating substructure elements is given in AASHTO Table 3.8.1.2.2-1 as 0.050 kips/ft². (The minimum load of 0.30 kips/ft from AASHTO Sec. 3.8.1.2.1 is only for analysis of superstructure elements and does not apply to the analysis of substructure elements impacted by wind loads on the superstructure.)

The base wind pressure applied directly to the substructure, $P_{B,sub}$, is given in AASHTO Sec. 3.8.1.2.3 as 0.040 kips/ft².

For a design wind velocity, v_{DZ}, greater than 100 mi/hr, the base pressures must be modified per AASHTO Eq. 3.8.1.2.1-1 to obtain the design wind pressures, $P_{D,super}$ and $P_{D,sub}$, for superstructures and

substructures, respectively. (AASHTO uses V for design wind velocity, whereas this solution uses v.)

$$P_{D,super} = P_{B,super}\left(\frac{v_{DZ}^2}{10,000}\right)$$
$$= \left(0.050 \ \frac{kips}{ft^2}\right)\left(\frac{\left(120 \ \frac{mi}{hr}\right)^2}{10,000 \ \frac{mi^2}{hr^2}}\right)$$
$$= 0.072 \ kips/ft^2$$

$$P_{D,sub} = P_{B,sub}\left(\frac{v_{DZ}^2}{10,000}\right) = \left(0.040 \ \frac{kips}{ft^2}\right)\left(\frac{\left(120 \ \frac{mi}{hr}\right)^2}{10,000 \ \frac{mi^2}{hr^2}}\right)$$
$$= 0.0576 \ kips/ft^2$$

Find the unfactored reaction, R, at the base of one of the interior piers by multiplying the design wind pressures by the surface areas of the superstructure and substructure elements tributary to an interior pier. D is the superstructure depth, L is the superstructure span, h is the pier height, and B is the pier width.

$$R = 2P_{D,super}D\left(\frac{L}{2}\right) + hBP_{D,sub}$$
$$= (2)\left(0.072 \ \frac{kips}{ft^2}\right)(5 \ ft)\left(\frac{100 \ ft}{2}\right)$$
$$\quad + (20 \ ft)(5 \ ft)\left(0.0576 \ \frac{kips}{ft^2}\right)$$
$$= 41.8 \ kips \quad (42 \ kips)$$

The answer is (C).

61. From AASHTO Sec. 3.8.1.3, the design wind pressure on vehicles, $P_{D,WL}$, is 0.10 kips/ft. This load acts 6 ft above the roadway.

The unfactored transverse moment, M, at the base of the pier is the design wind pressure on vehicles multiplied by length of bridge tributary to the pier and then multiplied by the height of the wind load above the base of the pier. L is the bridge span, h is the pier height, and t is the deck thickness.

$$M = 2P_{D,WL}\left(\frac{L}{2}\right)(h + t + 6 \text{ ft})$$

$$= (2)\left(0.10 \frac{\text{kips}}{\text{ft}}\right)\left(\frac{120 \text{ ft}}{2}\right)(20 \text{ ft} + 3 \text{ ft} + 6 \text{ ft})$$

$$= 348 \text{ ft-kips} \quad (350 \text{ ft-kips})$$

The answer is (C).

62. Using the portal method of simplified analysis, points of zero moment are assumed at the midpoints of all columns and beams. In addition, the shear at interior columns is assumed to be twice that of exterior columns.

Cut a free body diagram at the midpoint of the first story, and solve for the column shears, V.

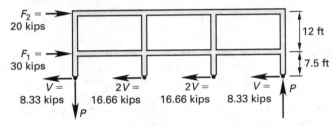

$$V + 2V + 2V + V = F_1 + F_2$$

$$V = \frac{F_1 + F_2}{6} = \frac{30 \text{ kips} + 20 \text{ kips}}{6} = 8.33 \text{ kips}$$

The shear in an interior column, V_{int}, is

$$V_{\text{int}} = 2V = (2)(8.33 \text{ kips}) = 16.66 \text{ kips}$$

The moment reaction at the base of an interior column, M_{int}, is

$$M_{\text{int}} = V_{\text{int}}\left(\frac{h_1}{2}\right) = (16.66 \text{ kips})\left(\frac{15 \text{ ft}}{2}\right)$$

$$= 125 \text{ ft-kips} \quad (130 \text{ ft-kips})$$

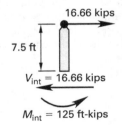

The answer is (C).

63. Story drift, Δ_x, is determined in accordance with ASCE/SEI7 Sec. 12.8.6 and Fig. 12.8-2.

$$\Delta_x = \frac{C_d(\delta_{x+1,e} - \delta_{xe})}{I}$$

From ASCE/SEI7 Table 1.5-2, the importance factor, I_e, for risk category IV buildings is 1.50.

The deflection amplification factor, C_d, is determined from ASCE/SEI7 Table 12.2-1. For steel special concentrically braced frames (ASCE/SEI7 Table 12.2-1, case B.2), C_d is 5.

Find the computed story drift at the tenth story using the provided elastic deflections at the ninth and tenth floors.

$$\Delta_{10} = \frac{C_d(\delta_{10} - \delta_9)}{I} = \frac{(5)(3.21 \text{ in} - 2.90 \text{ in})}{1.5} = 1.03 \text{ in}$$

Allowable story drifts, Δ_a, are given in ASCE/SEI7 Table 12.12-1. For "all other structures" under risk category IV,

$$\Delta_a = 0.010h_{sx} = (0.010)(12 \text{ ft})\left(12 \frac{\text{in}}{\text{ft}}\right) = 1.44 \text{ in}$$

The ratio of the computed story drift to the allowable story drift is

$$\frac{\Delta_{10}}{\Delta_a} = \frac{1.03 \text{ in}}{1.44 \text{ in}} = 0.72$$

The answer is (B).

64. By inspection, the controlling load combination for LRFD is from IBC Eq. 16-6.

$$0.9 + W$$

The diaphragm load due to wind pressure is equal to the wind pressure, q_{wind}, multiplied by half the building height, H. The lateral load, W, at the top of the shear wall is equal to the diaphragm load multiplied by the width of the west diaphragm tributary to wall 1.

$$W = q_{\text{wind}}\left(\frac{H}{2}\right)\left(\frac{L}{2}\right) = \left(40 \frac{\text{lbf}}{\text{ft}^2}\right)\left(\frac{12 \text{ ft}}{2}\right)\left(\frac{40 \text{ ft}}{2}\right)$$

$$= 4800 \text{ lbf}$$

Considering the framing that spans between shear wall 1 and the collector beam, the vertical dead load, D, on the wall is

$$D = q_{\text{dead}}B\left(\frac{L}{2}\right) = \left(15 \frac{\text{lbf}}{\text{ft}^2}\right)(20 \text{ ft})\left(\frac{40 \text{ ft}}{2}\right)$$

$$= 6000 \text{ lbf}$$

(If the self-weight of the wall had been given, this could have been included as part of the dead load, D.)

Taking moments about one of the tie-downs, the overturning moment, M, acting on the wall is

$$M = WH - 0.9\left(\frac{b}{2}\right)$$
$$= (4800 \text{ lbf})(12 \text{ ft}) - (0.9)(6000 \text{ lbf})\left(\frac{18 \text{ ft}}{2}\right)$$
$$= 9000 \text{ ft-lbf}$$

The net uplift on one of the tie-downs, T, is the overturning moment divided by the tie-down spacing, b.

$$T = \frac{M}{b} = \frac{9000 \text{ ft-lbf}}{18 \text{ ft}} = 500 \text{ lbf}$$

The answer is (B).

65. The required anchorage design force, F_p, for structural walls connected to flexible diaphragms is given in ASCE/SEI7 Sec. 12.11.2.1. F_p is found from ASCE/SEI7 Eq. 12.11-1.

$$F_p = 0.4 S_{DS} K_a I_e W_p$$

From ASCE/SEI7 Eq. 12.11-2,

$$K_a = 1.0 + \frac{L_f}{100} = 1.0 + \frac{100}{100}$$
$$= 2.0$$

Considering the top half of the wall plus the parapet tributary to the anchor, the weight of the wall tributary to the anchor, W_p, is

$$W_p = \gamma_c t s\left(\frac{h}{2} + a\right)$$
$$= \left(150 \frac{\text{lbf}}{\text{ft}^3}\right)\left(\frac{8 \text{ in}}{12 \frac{\text{in}}{\text{ft}}}\right)(4 \text{ ft})\left(\frac{12 \text{ ft}}{2} + 3.5 \text{ ft}\right)$$
$$= 3800 \text{ lbf}$$

The out-of-plane force, F_p, is

$$F_p = 0.4 S_{DS} K_a I_e W_p = (0.4)(0.60)(2.0)(1.0)(3800 \text{ lbf})$$
$$= 1824 \text{ lbf}$$

From ASCE/SEI7 Sec. 12.11.2.1, F_p cannot be taken as less than

$$F_p = 0.2 K_a I_e W_p = (0.2)(2.0)(1.0)(3800 \text{ lbf})$$
$$= 1520 \text{ lbf} \quad [\text{does not control}]$$

Therefore, the controlling anchorage design force is $F_p = 1824 \text{ lbf}$.

From ASCE/SEI7 Sec. 12.11.2.2.2, the strength design forces for steel elements of the structural wall anchorage system must be increased by a factor of 1.4. From IBC Sec. 1605.2.1, the load factor for earthquake loads is 1.0 for strength design. The strength design tensile force in the steel strap, T, is

$$T = (1.4)(1.0) F_p = (1.4)(1.0)(1824 \text{ lbf})$$
$$= 2554 \text{ lbf} \quad (2600 \text{ lbf})$$

The answer is (D).

66. Minimum support length requirements are given in AASHTO Sec. 4.7.4.4.

From the problem illustration, the length, L, from the end of bridge deck to the adjacent expansion joint is 80 ft, and the average height, H, of the bridge columns is 17.5 ft. The skew of bridge supports, S, is 0° because the bridge is not skewed.

From AASHTO Table 4.7.4.4-1, for a bridge in seismic zone 3, the minimum support length is 150% of the support length, N.

The minimum support length is calculated from AASHTO Eq. 4.7.4.4-1, adjusted for the minimum support length percentage of N. (This equation is not dimensionally consistent.)

$$N = 1.5(8 + 0.02L + 0.08H)(1 + 0.000125 S^2)$$
$$= (1.5)(8 + (0.02)(80 \text{ ft}) + (0.08)(17.5 \text{ ft}))$$
$$\times (1 + (0.000125)(0°)^2)$$
$$= 16.5 \text{ in} \quad (17 \text{ in})$$

The answer is (D).

67. The AISC *Seismic Construction Manual*, Part 5.4, describes eccentrically braced frames (EBFs) as frames where one end of each brace intersects a beam at an eccentricity, forming a link that is subject to shear and flexure. The link is the primary location of inelastic behavior in the frame. The remainder of the members and connections are intended to remain essentially elastic, so option A is incorrect.

EBFs can often fit in locations within the architectural floor plan where concentrically braced frames cannot due to the space limitations that doors and windows present. EBFs also have higher response modification factors than concentrically braced frames (7 or 8 versus 6, respectively), and therefore, lower seismic base shears than concentrically braced frames. EBFs also provide for greater lateral stiffness than moment frames, thereby limiting nonstructural damage.

The answer is (A).

68. ASD Solution

AISC 360 Sec. D2 states that the allowable tensile strength is the lower value obtained when considering tensile yielding in the gross section and tensile rupture in the net section. Per AISC 360 Sec. D3, the gross area and net area are determined using AISC 360 Sec. B4.3.

Per AISC 360 Sec. B4.3a, the gross area, A_g, of the member is

$$A_g = tb = \left(\frac{1}{2} \text{ in}\right)(6 \text{ in}) = 3.0 \text{ in}^2$$

From AISC 360 Eq. D2-1, the allowable tensile yielding strength is

$$\frac{P_n}{\Omega_t} = \frac{F_y A_g}{\Omega_t} = \frac{\left(42 \ \frac{\text{kips}}{\text{in}^2}\right)(3.0 \text{ in}^2)}{1.67}$$
$$= 75.4 \text{ kips} \quad [\text{use 75 kips}]$$

For tensile rupture, AISC 360 Table J3.3 gives the diameter of a standard hole for a ¾ in diameter bolt as $^{13}\!/_{16}$ in. Per AISC 360 Sec. B4.3b, the width of the bolt hole, d_h, used to calculate the net area is the nominal dimension of the hole plus $^1\!/_{16}$ in.

$$d_h = \frac{13}{16} \text{ in} + \frac{1}{16} \text{ in} = \frac{7}{8} \text{ in}$$

From AISC 360 Sec. B4.3b, the net area, A_n, is the lesser of

$$A_n = \min \begin{cases} t(b - d_h) = \left(\frac{1}{2} \text{ in}\right)\left(6 \text{ in} - \frac{7}{8} \text{ in}\right) \\ \qquad = 2.56 \text{ in}^2 \\ t\left(b - 2d_h + \frac{s^2}{4g}\right) = \left(\frac{1}{2} \text{ in}\right)\begin{pmatrix} 6 \text{ in} \\ -(2)\left(\frac{7}{8} \text{ in}\right) \\ +\frac{(1.5 \text{ in})^2}{(4)(3 \text{ in})} \end{pmatrix} \\ \qquad = 2.22 \text{ in}^2 \quad [\text{controls}] \end{cases}$$

From AISC 360 Table D3.1, case 1, the shear lag factor, U, is 1.0. From AISC 360 Eq. D3-1, the effective area, A_e, of the tension member is then

$$A_e = A_n U = (2.22 \text{ in}^2)(1.0) = 2.22 \text{ in}^2$$

Per AISC 360 Sec. J4.1(b), A_e must be less than $0.85 A_g$.

$$A_e < 0.85 A_g$$
$$2.22 \text{ in}^2 < (0.85)(3.0 \text{ in}^2)$$
$$2.22 \text{ in}^2 < 2.55 \text{ in}^2 \quad [\text{OK}]$$

From AISC 360 Eq. D2-2, the allowable tensile rupture strength is

$$\frac{P_n}{\Omega_t} = \frac{F_u A_e}{\Omega_t} = \frac{\left(60 \ \frac{\text{kips}}{\text{in}^2}\right)(2.22 \text{ in}^2)}{2.00}$$
$$= 66.6 \text{ kips} \quad (67 \text{ kips})$$

67 kips < 75 kips; therefore, tensile rupture controls over tensile yielding.

The answer is (B).

LRFD Solution

AISC 360 Sec. D2 states that the design tensile strength is the lower value obtained when considering tensile yielding in the gross section and tensile rupture in the net section. Per AISC Sec. D3, the gross area and net area are determined using AISC 360 Sec. B4.3.

Per AISC 360 Sec. B4.3a, the gross area, A_g, of the member is

$$A_g = tb = \left(\frac{1}{2} \text{ in}\right)(6 \text{ in}) = 3.0 \text{ in}^2$$

From AISC 360 Eq. D2-1, the design tensile yielding strength is

$$\phi_t P_n = \phi_t F_y A_g = (0.90)\left(42 \ \frac{\text{kips}}{\text{in}^2}\right)(3.0 \text{ in}^2)$$
$$= 113 \text{ kips}$$

For tensile rupture, AISC 360 Table J3.3 gives the diameter of a standard hole for a ¾ in diameter bolt as $^{13}\!/_{16}$ in. Per AISC 360 Sec. B4.3b, the width of the bolt hole, d_h, used to calculate the net area is the nominal dimension of the hole plus $^1\!/_{16}$ in.

$$d_h = \frac{13}{16} \text{ in} + \frac{1}{16} \text{ in} = \frac{7}{8} \text{ in}$$

Per AISC 360 Sec. B4.3b, the net area, A_n, is the lesser of

$$A_n = \min \begin{cases} t(b - d_h) = \left(\dfrac{1}{2} \text{ in}\right)\left(6 \text{ in} - \dfrac{7}{8} \text{ in}\right) \\ \qquad = 2.56 \text{ in}^2 \\ t\left(b - 2d_h + \dfrac{s^2}{4g}\right) = \left(\dfrac{1}{2} \text{ in}\right)\begin{pmatrix} 6 \text{ in} \\ -(2)\left(\dfrac{7}{8} \text{ in}\right) \\ +\dfrac{(1.5 \text{ in})^2}{(4)(3 \text{ in})} \end{pmatrix} \\ \qquad = 2.22 \text{ in}^2 \quad \text{[controls]} \end{cases}$$

From AISC 360 Table D3.1, case 1, the shear lag factor, U, is 1.0. From AISC 360 Eq. D3-1, the effective area, A_e, of the tension member is then

$$A_e = A_n U = (2.22 \text{ in}^2)(1.0) = 2.22 \text{ in}^2$$

Per AISC 360 Sec. J4.1(b), A_e must be less than $0.85A_g$.

$$A_e < 0.85A_g$$
$$2.22 \text{ in}^2 < (0.85)(3.0 \text{ in}^2)$$
$$2.22 \text{ in}^2 < 2.55 \text{ in}^2 \quad \text{[OK]}$$

From AISC 360 Eq. D2-2, the design tensile rupture strength is

$$\phi_t P_n = \phi_t F_u A_e = (0.75)\left(60 \ \frac{\text{kips}}{\text{in}^2}\right)(2.22 \text{ in}^2)$$
$$= 100 \text{ kips}$$

100 kips < 113 kips; therefore, tensile rupture controls over tensile yielding.

The answer is (B).

69. The design procedure for reduced beam section (RBS) moment connections is described in AISC 358 (Part 9.2 of the AISC *Seismic Construction Manual*) Sec. 5.8. The procedure is also outlined in the AISC *Seismic Construction Manual* Ex. 4.34.

From AISC 358 Eq. 5.8-4, at the centerline of the reduced beam section, the plastic section modulus, Z_{RBS}, is

$$Z_{\text{RBS}} = Z_e = Z_x - 2ct_f(d - t_f)$$
$$= 144 \text{ in}^3 - (2)(1.5 \text{ in})(0.615 \text{ in})(21 \text{ in} - 0.615 \text{ in})$$
$$= 106.4 \text{ in}^3$$

From AISC 341 (Part 9.1 of the AISC *Seismic Construction Manual*) Table A3.1, for ASTM A992 grade 55

steel, the ratio of expected yield stress to specified minimum yield stress, R_y, is 1.1.

From AISC 358 Eq. 2.4.3-2, the peak connection strength factor, C_{pr}, is

$$C_{pr} = \frac{F_y + F_u}{2F_y} \leq 1.2$$
$$= \frac{55 \ \dfrac{\text{kips}}{\text{in}^2} + 65 \ \dfrac{\text{kips}}{\text{in}^2}}{(2)\left(55 \ \dfrac{\text{kips}}{\text{in}^2}\right)}$$
$$= 1.09 \quad [\leq 1.2, \text{ OK}]$$

From AISC 358 Eq. 2.4.3-1 (or Eq. 5.8-5), the probable maximum moment, M_{pr}, at the center of the RBS is

$$M_{pr} = C_{pr} R_y F_y Z_e$$
$$= \frac{(1.09)(1.1)\left(55 \ \dfrac{\text{kips}}{\text{in}^2}\right)(106.4 \text{ in}^3)}{12 \ \dfrac{\text{in}}{\text{ft}}}$$
$$= 585 \text{ ft-kips} \quad (590 \text{ ft-kips})$$

The answer is (B).

70. Per AISC 341 (Part 9.1 of the AISC *Seismic Construction Manual*) Sec. F3.5b(2), the link's shear strength is the lower value obtained when considering the limit states of shear yielding in the web and flexural yielding in the gross section.

The effect of the axial force on the link can be ignored if

$$\frac{P_r}{P_c} = \frac{P_u}{P_y} \leq 0.15$$

$$P_y = F_y A_g = \left(50 \ \frac{\text{kips}}{\text{in}^2}\right)(20.8 \text{ in}^3)$$
$$= 1040 \text{ kips}$$

$$\frac{150 \text{ kips}}{1040 \text{ kips}} = 0.144 \leq 0.15 \quad \text{[ignore axial force]}$$

For shear yielding,

$$v_n = v_p = 0.6 F_y A_{lw}$$
$$= (0.6)\left(50 \ \frac{\text{kips}}{\text{in}^2}\right)(8.36 \text{ in}^2)$$
$$= 251 \text{ kips} \quad (250 \text{ kips}) \quad \text{[controls]}$$

For flexural yielding,

$$M_p = F_y Z = \left(50 \ \frac{\text{kips}}{\text{in}^2}\right)(146 \text{ in}^3)$$

$$= 7300 \text{ in-kips}$$

$$V_n = \frac{2M_p}{e} = \frac{(2)(7300 \text{ in-kips})}{42 \text{ in}}$$

$$= 348 \text{ kips} \quad [\text{does not control}]$$

The answer is (B).

71. The maximum diaphragm moment, M, is

$$M = \frac{wL^2}{8} = \frac{\left(300 \ \frac{\text{lbf}}{\text{ft}}\right)(50 \text{ ft})^2}{8} = 93{,}750 \text{ ft-lbf}$$

The maximum chord force, T, is the diaphragm moment divided by the distance between the diaphragm straps, B.

$$T = \frac{M}{B} = \frac{93{,}750 \text{ ft-lbf}}{20 \text{ ft}} = 4688 \text{ lbf}$$

The thickness of 18-gage straps is given as 0.048 in, so $t_1 = t_2 = 0.048$ in. *AISI* Sec. E4.3.1 states that when $t_2/t_1 \leq 1.0$, the nominal shear strength per screw, P_{ns}, is the smaller of

$$P_{ns} = \min \begin{cases} 4.2(t^3 d)^{1/2} F_u = (4.2)\begin{pmatrix}(0.048 \text{ in})^3 \\ \times (0.190 \text{ in})\end{pmatrix}^{1/2} \\ \qquad \times \left(65{,}000 \ \frac{\text{lbf}}{\text{in}^2}\right) \\ \qquad = 1251 \text{ lbf} \quad [\text{controls}] \\ 2.7 t d F_u = (2.7)(0.048 \text{ in})(0.190 \text{ in}) \\ \qquad \times \left(65{,}000 \ \frac{\text{lbf}}{\text{in}^2}\right) \\ \qquad = 1601 \text{ lbf} \end{cases}$$

(The controlling nominal shear strength could also have been found in *AISI Manual* Table IV-9b as $P_{ns} = 1250$ lbf.)

The required number of screws, n_{screws}, is

$$n_{\text{screws}} = \frac{T}{\dfrac{P_{ns}}{\Omega}} = \frac{4688 \text{ lbf}}{\dfrac{1250 \text{ lbf}}{3.0}} = 11.3 \quad (12 \text{ screws})$$

The answer is (D).

72. From ACI 318 Eq. 18.10.7.4, the nominal shear capacity, V_n, of coupling beams reinforced with two intersecting groups of diagonal bars is the lesser of

$$V_n = \min \begin{cases} 2A_{vd}f_y \sin \alpha = (2)(8 \text{ bars})\left(0.20 \ \frac{\text{in}^2}{\text{bar}}\right) \\ \qquad \times \left(60 \ \frac{\text{kips}}{\text{in}^2}\right)\sin 30° \\ \qquad = 96 \text{ kips} \quad [\text{controls}] \\ 10\sqrt{f_c'} A_{cw} = \dfrac{(10)\sqrt{4000 \ \dfrac{\text{lbf}}{\text{in}^2}}(18 \text{ in})(36 \text{ in})}{1000 \ \dfrac{\text{lbf}}{\text{kip}}} \\ \qquad = 410 \text{ kips} \end{cases}$$

The answer is (B).

73. For intermediate concrete moment frames, the design shear force, V_u, is a function of the nominal moment capacity of the beam.

The depth of the equivalent rectangular stress block, a, is

$$a = \frac{A_s f_y}{0.85 f_c' b} = \frac{(3.0 \text{ in}^2)\left(60 \ \frac{\text{kips}}{\text{in}^2}\right)\left(1000 \ \frac{\text{lbf}}{\text{kip}}\right)}{(0.85)\left(4000 \ \frac{\text{lbf}}{\text{in}^2}\right)(12 \text{ in})}$$

$$= 4.41 \text{ in}$$

The nominal moment capacity at the left and right sides of the beam, $M_{nl} = M_{nr}$, is

$$M_{nl} = M_{nr} = A_s f_y \left(d - \frac{a}{2}\right)$$

$$= \frac{(3.0 \text{ in}^2)\left(60 \ \frac{\text{kips}}{\text{in}^2}\right)\left(16 \text{ in} - \frac{4.41 \text{ in}}{2}\right)}{12 \ \frac{\text{in}}{\text{ft}}}$$

$$= 207 \text{ ft-kips}$$

From ACI 318 Eq. 5.3.1e, the factored distributed beam load, w_u, is

$$w_u = 1.2 w_D + 1.0 w_L$$

$$= (1.2)\left(1.5 \ \frac{\text{kips}}{\text{ft}}\right) + (1.0)\left(0.8 \ \frac{\text{kip}}{\text{ft}}\right)$$

$$= 2.6 \text{ kips/ft}$$

The factored shear force for the beams and columns of intermediate moment frames is determined in accordance with ACI 318 Sec. 18.4.2.3. For a beam resisting earthquake loads, the design shear force, V_u, is the smaller of (a) or (b).

(a) the sum of the shear associated with development of nominal moment strengths of the beam at each restrained end of the clear span due to reverse curvature bending and the shear calculated for factored gravity loads

From ACI 318 R18.4.2,

$$V_u = \frac{M_{nl} + M_{nr}}{l_n} + \frac{w_u l_n}{2}$$

$$= \frac{207 \text{ ft-kips} + 207 \text{ ft-kips}}{20 \text{ ft}} + \frac{\left(2.6 \dfrac{\text{kips}}{\text{ft}}\right)(20 \text{ ft})}{2}$$

$$= 47 \text{ kips} \quad [\text{does not control}]$$

(b) the maximum shear obtained from design load combinations that include E, with E assumed to be twice that prescribed by the legally adopted general building code for earthquake-resistant design

$$V_u = \frac{w_u l_n}{2} + 2.0 V_{EQ}$$

$$= \frac{\left(2.6 \dfrac{\text{kips}}{\text{ft}}\right)(20 \text{ ft})}{2} + (2.0)(8.0 \text{ kips})$$

$$= 42 \text{ kips} \quad [\text{controls}]$$

The answer is (C).

74. Determine if slenderness effects may be neglected per ACI 318 Sec. 6.2.5. From ACI 318 Sec. 10.10.1.2, the radius of gyration, r, is 0.30 times the overall thickness, h, for rectangular compression members.

$$r = 0.3h = \frac{(0.3)(14 \text{ in})}{12 \dfrac{\text{in}}{\text{ft}}} = 0.35 \text{ ft}$$

From ACI 318 Sec. 6.2.5(b), slenderness effects may be neglected for compression members braced against sidesway when ACI 318 Eq. 6.2.5b and 6.2.5c are is satisfied.

$$\frac{kl_u}{r} \leq 34 + 12\left(\frac{M_1}{M_2}\right)$$

$$\text{and } \frac{kl_u}{r} \leq 40$$

The shown moments result in double curvature of the column. In accordance with the definitions for M_1 and

M_2 from ACI 318 Sec. 2.2, take M_1 as -10 ft-kips and M_2 as 65 ft-kips.

$$\frac{18 \text{ ft}}{0.35 \text{ ft}} \leq 34 + (12)\left(\frac{10 \text{ ft-kips}}{65 \text{ ft-kips}}\right)$$

Since $51.4 > 35.8$ slenderness effects cannot be neglected.

From ACI 318 Sec. 6.2.6, the design of members must be based on a second-order analysis. Nonlinear (see ACI 318 Sec. 6.8) and elastic (see ACI 318 Sec. 6.7) versions of second-order analysis are not possible with the limited information given in the problem statement. Therefore, use the moment magnification procedure for nonsway frames per ACI 318 Sec. 6.6.4.

EI is given in the design criteria as 2.885×10^6 in²-kips. From ACI 318 Eq. 6.6.4.4.2, the critical buckling load, P_c, is

$$P_c = \frac{\pi^2 EI}{(kl_u)^2} = \frac{\pi^2 (2.885 \times 10^6 \text{ in}^2\text{-kips})}{\left((18 \text{ ft})\left(12 \dfrac{\text{in}}{\text{ft}}\right)\right)^2}$$

$$= 610.3 \text{ kips}$$

From ACI 318 Eq. 6.6.4.5.3, the factor relating the actual moment diagram to an equivalent, uniform moment diagram, C_m, is

$$C_m = 0.6 + 0.4\left(\frac{M_1}{M_2}\right) = 0.6 - (0.4)\left(\frac{-10 \text{ ft-kips}}{65 \text{ ft-kips}}\right)$$

$$= 0.538$$

Check the minimum factored moment, $M_{2,\text{min}}$, using ACI 318 Sec. 6.6.4.5.4.

$$M_{2,\text{min}} = P_u(0.6 + 0.03h)$$

$$= \frac{(250 \text{ kips})(0.6 + (0.03)(14 \text{ in}))}{12 \dfrac{\text{in}}{\text{ft}}}$$

$$= 21.25 \text{ ft-kips} \leq 65 \text{ ft-kips} \quad [\text{OK}]$$

From ACI 318 Eq. 6.6.4.5.2, the moment magnification factor for nonsway frames, δ, is

$$\delta = \frac{C_m}{1 - \dfrac{P_u}{0.75 P_c}} = \frac{0.538}{1 - \dfrac{250 \text{ kips}}{(0.75)(610.3 \text{ kips})}}$$

$$= 1.185 \quad [\geq 1.0, \text{ OK}]$$

From ACI 318 Eq. 6.6.4.5.1, the column must be designed for a factored moment of

$$M_c = \delta M_2 = (1.185)(65 \text{ ft-kips}) = 77 \text{ ft-kips}$$

The answer is (B).

75. The distance from the center of the anchor to the edge of the concrete in the direction of the applied shear, c_{a1}, is 8 in. The thickness of the member where the anchor is located, h_a, is 10 in. From ACI 318 Eq. 17.5.2.1c, neglecting edge distances, the projected concrete failure area, A_{Vco}, is

$$A_{Vco} = 4.5 c_{a1}^2 = (4.5)(8 \text{ in})^2 = 288 \text{ in}^2$$

From ACI 318 Fig. R17.5.2.1b, for $h_a < 1.5 c_{a1}$, the concrete failure area, A_{Vc}, is

$$A_{Vc} = 2(1.5 c_{a1})h_a = (2)\big((1.5)(8 \text{ in})\big)(10 \text{ in})$$
$$= 240 \text{ in}^2$$

From ACI 318 Eq.17.5.2.8, for $h_a < 1.5 c_{a1}$, the modification factor, $\Psi_{h,V}$, is

$$\Psi_{h,V} = \sqrt{\frac{1.5 c_{a1}}{h_a}} = \sqrt{\frac{(1.5)(8 \text{ in})}{10 \text{ in}}} = 1.095$$

For a single anchor, the nominal concrete breakout strength, V_{cb}, for a shear force perpendicular to the edge is given by ACI 318 Eq. 17.5.2.1a.

$$V_n = V_{cb} = \frac{A_{Vc}}{A_{Vco}} \Psi_{ed,V} \Psi_{c,V} \Psi_{h,V} V_b$$
$$= \left(\frac{240 \text{ in}^2}{288 \text{ in}^2}\right)(1.0)(1.0)(1.095)(11.5 \text{ kips})$$
$$= 10.5 \text{ kips}$$

The answer is (A).

76. The minimum volumetric spiral reinforcement ratio is given in AASHTO Sec. 5.7.4.6. For columns in seismic zones 2, 3, or 4, the area of spiral reinforcement must also satisfy AASHTO Sec. 5.10.11.4.1d.

The gross area, A_g, of the column is

$$A_g = \frac{\pi D^2}{4} = \frac{\pi (36 \text{ in})^2}{4} = 1018 \text{ in}^2$$

Find the area of the core, A_c, measured from the outside of the spirals. c_c is the clear cover of the reinforcement.

$$A_c = \frac{\pi (D - 2 c_c)^2}{4} = \frac{\pi\big(36 \text{ in} - (2)(1.5 \text{ in})\big)^2}{4} = 855 \text{ in}^2$$

From AASHTO Eq. 5.7.4.6-1, the required spiral reinforcement ratio, ρ_s, is

$$\rho_s \geq 0.45 \left(\frac{A_g}{A_c} - 1\right)\left(\frac{f_c'}{f_{yh}}\right)$$
$$= (0.45)\left(\frac{1018 \text{ in}^2}{855 \text{ in}^2} - 1\right)\left(\frac{4 \dfrac{\text{kips}}{\text{in}^2}}{60 \dfrac{\text{kips}}{\text{in}^2}}\right)$$
$$= 0.00572 \quad [\text{does not control}]$$

From AASHTO Sec. 5.10.11.4.1d, to meet seismic detailing requirements, the spiral reinforcement ratio must satisfy either that required by AASHTO Sec. 5.7.4.6 or

$$\rho_s \geq 0.12\left(\frac{f_c'}{f_y}\right) = (0.12)\left(\frac{4 \dfrac{\text{kips}}{\text{in}^2}}{60 \dfrac{\text{kips}}{\text{in}^2}}\right)$$
$$= 0.008 \quad [\text{does not control}]$$

The spiral reinforcement ratio is defined as the reinforcing steel volume divided by the concrete volume in the core for one turn of the spiral bar. s is the pitch or spacing of the spirals, l is the length of one turn of the spiral, and A_s is the cross-sectional area of the spiral.

$$\rho_s = \frac{\text{steel volume in one turn}}{\text{concrete volume in one turn}} = \frac{A_s l}{A_c s}$$

Set the spiral reinforcement ratio, ρ_s, equal to 0.00572 and solve for the pitch, s. The area, A_s, of a no. 3 bar is 0.11 in² per ACI 318 App. E.

$$s \leq \frac{A_s l}{\rho_s A_c} = \frac{(0.11 \text{ in}^2)\big(36 \text{ in} - (2)(1.5 \text{ in})\big)\pi}{(0.00572)(855 \text{ in}^2)}$$
$$= 2.33 \text{ in} \quad (2.0 \text{ in}) \quad [\text{controls}]$$

From AASHTO Sec. 5.10.6.2, where d_b is the longitudinal bar diameter, the minimum spacing, $s_{\min}$, is

$$s_{\min} = 6 d_b = (6)(1.41 \text{ in}) = 8.46 \text{ in} \quad [\text{does not control}]$$

The answer is (A).

77. The allowable diaphragm shear can be determined from SDPWS Table 4.2C. The loading direction and panel configuration correspond to loading case 5.

From SDPWS Table 4.2C, a case 5, unblocked diaphragm with structural I grade panels fastened to 3 in nominal width members with 10d common wire nails has a nominal unit shear capacity, v_w, of 670 lbf/ft for wind loading.

From SDPWS Table 4.2C, ftn. 2, this value must be adjusted by the specific gravity of spruce-pine-fir. From NDS Table 11.3.2A, the specific gravity, SG, for spruce-pine-fir is 0.42.

From SDPWS Sec. 4.2.3, for ASD, the nominal unit shear capacities must be divided by 2.0. Therefore, the adjusted allowable diaphragm shear, v'_w, for allowable stress design is

$$
\begin{aligned}
v'_w &= \frac{\left(670 \ \frac{\text{lbf}}{\text{ft}}\right)\left(1-(0.5-\text{SG})\right)}{2.0} \\
&= \frac{\left(670 \ \frac{\text{lbf}}{\text{ft}}\right)\left(1-(0.5-0.42)\right)}{2.0} \\
&= 308 \ \text{lbf/ft} \quad (310 \ \text{lbf/ft})
\end{aligned}
$$

The answer is (B).

78. For a flexible diaphragm, each line of shear walls will resist loads in proportion to their tributary widths.

The diaphragm reaction, R, is

$$
R = \frac{q_{\text{wind}} L}{2} = \frac{\left(200 \ \frac{\text{lbf}}{\text{ft}}\right)(70 \ \text{ft})}{2} = 7000 \ \text{lbf}
$$

The unit shear of the diaphragm reaction, v_{dia}, is the diaphragm reaction divided by the length of the diaphragm.

$$
v_{\text{dia}} = \frac{R}{a+b+c} = \frac{7000 \ \text{lbf}}{20 \ \text{ft}+50 \ \text{ft}+30 \ \text{ft}} = 70 \ \text{lbf/ft}
$$

The unit shear in the shear walls, v_{wall}, on either end of the collector is the diaphragm reaction divided by the length of the shear walls.

$$
v_{\text{wall}} = \frac{R}{a+c} = \frac{7000 \ \text{lbf}}{20 \ \text{ft}+30 \ \text{ft}} = 140 \ \text{lbf/ft}
$$

The axial force at the left end of the collector, P_l, is

$$
\begin{aligned}
P_l &= (v_{\text{wall}} - v_{\text{dia}})a = \left(140 \ \frac{\text{lbf}}{\text{ft}} - 70 \ \frac{\text{lbf}}{\text{ft}}\right)(20 \ \text{ft}) \\
&= 1400 \ \text{lbf} \quad [\text{does not control}]
\end{aligned}
$$

The compression force at the right end of the collector, P_r, is

$$
\begin{aligned}
P_r &= (v_{\text{dia}} - v_{\text{wall}})c = \left(70 \ \frac{\text{lbf}}{\text{ft}} - 140 \ \frac{\text{lbf}}{\text{ft}}\right)(30 \ \text{ft}) \\
&= -2100 \ \text{lbf} \quad [\text{controls}]
\end{aligned}
$$

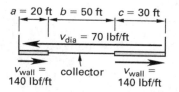

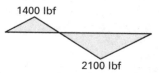

collector axial force diagram

The answer is (B).

79. The factored moment at the mid-height of the wall is determined using TMS 402 Sec. 9.3.5.4.2.

The eccentricity of the load from the tributary roof area, e_u, is given in the problem illustration as 2.5 in.

From TMS 402 Eq. 9-27,

$$
\begin{aligned}
M_u &= \frac{w_u h^2}{8} + P_{uf}\frac{e_u}{2} + P_u\delta \\
&= \frac{\left(45 \ \frac{\text{lbf}}{\text{ft}^2}\right)(12 \ \text{ft})^2}{8} + \left(1600 \ \frac{\text{lbf}}{\text{ft}}\right)\left(\frac{2.5 \ \text{in}}{(2)\left(12 \ \frac{\text{in}}{\text{ft}}\right)}\right) \\
&\quad + \left(2300 \ \frac{\text{lbf}}{\text{ft}}\right)\left(\frac{0.08 \ \text{in}}{12 \ \frac{\text{in}}{\text{ft}}}\right) \\
&= 992 \ \text{ft-lbf/ft} \quad (990 \ \text{ft-lbf/ft})
\end{aligned}
$$

The answer is (D).

80. The detailing requirements for special reinforced masonry shear walls are given in TMS 402 Sec. 7.3.2.6. From TMS 402 Sec. 7.3.2.6(a), the maximum spacing of vertical reinforcement, s_{max}, is the smallest of 48 in, one-third the length of the shear wall, or one-third the height of the shear wall.

$$s_{max} = \min \begin{cases} 48 \text{ in} \\ \dfrac{L}{3} = \dfrac{(10 \text{ ft})\left(12 \, \dfrac{\text{in}}{\text{ft}}\right)}{3} = 40 \text{ in} \quad \text{[controls]} \\ \dfrac{H}{3} = \dfrac{(12 \text{ ft})\left(12 \, \dfrac{\text{in}}{\text{ft}}\right)}{3} = 48 \text{ in} \end{cases}$$

From TMS 402 Sec. 7.3.2.6(c), the minimum cross-sectional area of vertical reinforcement must be one-third of the cross-sectional area of the required shear reinforcement. The nominal area of no. 4 bars is 0.20 in^2 per ACI 318 App. A. Find the minimum required area of vertical shear reinforcement, $A_{sv,min}$. Assume the area of the provided horizontal bars is equal to the required area of shear reinforcement, A_v.

$$A_{sv,min} = \frac{A_v}{3} = \frac{(0.20 \text{ in}^2)(9 \text{ bars})}{3} = 0.60 \text{ in}^2$$

The nominal area of no. 8 bars is 0.79 in^2 per ACI 318 App. A. The cross-sectional area of the two vertical no. 8 flexural reinforcing bars is

$$A_{sv,provided} = (0.79 \text{ in}^2)(2 \text{ bars})$$
$$= 1.58 \text{ in}^2 > 0.60 \text{ in}^2 \quad \text{[OK]}$$

The two vertical no. 8 flexural reinforcing bars alone satisfy the requirements of TMS 402 Sec. 7.3.2.6(c), so this requirement will not govern the spacing of additional vertical bars.

From TMS 402 Sec. 7.3.2.6(c)(1), for masonry laid in running bond, the minimum cross-sectional area of reinforcement in each direction cannot not be less than 0.0007 multiplied by the gross cross-sectional area of the wall. The thickness of a nominal 10 in wall is 9.625 in. Therefore, the minimum required area of vertical shear reinforcement is

$$A_{sv,min} = 0.0007 A_g$$
$$= (0.0007)(9.625 \text{ in})(10 \text{ ft})\left(12 \, \frac{\text{in}}{\text{ft}}\right)$$
$$= 0.81 \text{ in}^2$$

This is less than the cross-sectional area of the two vertical no. 8 flexural reinforcing bars alone.

$$A_{sv,provided} = (0.79 \text{ in}^2)(2 \text{ bars})$$
$$= 1.58 \text{ in}^2 > 0.81 \text{ in}^2 \quad \text{[OK]}$$

Therefore, the 40 in spacing requirement governs. Using two additional vertical bars gives a spacing, s, of

$$s = \frac{L - 2(\text{cover})}{\text{no. of spaces}} = \frac{(10 \text{ ft})\left(12 \, \dfrac{\text{in}}{\text{ft}}\right) - (2)(4 \text{ in})}{3 \text{ spaces}}$$
$$= 37.3 \text{ in} \leq 40 \text{ in} \quad \text{[OK]}$$

Therefore, two additional vertical no. 4 bars are adequate.

The answer is (A).

81. The allowable axial tensile load of headed anchor bolts, B_a, is given in TMS 402 Sec. 8.1.3.3.1.1. It is the minimum of the value governed by either masonry breakout or steel yielding.

From TMS 402 Sec. 6.2.4, the effective embedment length for headed anchor bolts, l_b, is the embedded length measured from the masonry surface to the compression-bearing surface. From the problem illustration, $l_b = 6$ in.

The embedment length is more than half the thickness of the wall; therefore, the projected tension area will be truncated, as shown in TMS 402 Comm. Fig. CC-6.2-5.

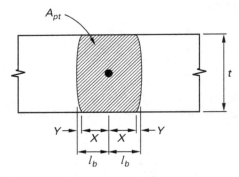

The value of A_{pt} is given in the problem statement as 64.9 in^2; however, it could have been calculated as follows.

From TMS 402 Comm. Fig. CC-6.2-5, the projected tension area for one anchor is

$$X = \frac{1}{2}\sqrt{4l_b^2 - t^2}$$

$$= \left(\frac{1}{2}\right)\sqrt{(4)(6 \text{ in})^2 - (5.625 \text{ in})^2}$$

$$= 5.30 \text{ in}$$

$$Y = l_b - X = 6 \text{ in} - 5.30 \text{ in}$$

$$= 0.70 \text{ in}$$

For the capacity of a single anchor, $Z = 0$ and the projected tension area on the masonry surface of a right circular cone, A_{pt}, is

$$A_{pt} = 2l_b t - Y\frac{2}{3}t$$

$$= (2)(6 \text{ in})(5.625 \text{ in}) - (0.70 \text{ in})\left(\frac{2}{3}\right)(5.625 \text{ in})$$

$$= 64.9 \text{ in}^2$$

From TMS 402 Eq. 8-1, the allowable tensile load for masonry breakout, B_{ab}, is

$$B_{ab} = 1.25 A_{pt}\sqrt{f'_m}$$

$$= \frac{(1.25)(64.9 \text{ in}^2)\sqrt{1500 \dfrac{\text{lbf}}{\text{in}^2}}}{1000 \dfrac{\text{lbf}}{\text{kip}}}$$

$$= 3.1 \text{ kips} \quad \text{[controls]}$$

From TMS 402 Eq. 8-2, the allowable tensile load for yielding of the anchor, B_{as}, is

$$B_{as} = 0.6 A_b f_y$$

$$= (0.6)\left(\frac{\pi\left(\frac{1}{2}\text{ in}\right)^2}{4}\right)\left(55 \frac{\text{kips}}{\text{in}^2}\right)$$

$$= 6.5 \text{ kips} \quad \text{[does not control]}$$

The answer is (B).

82. For a rigid pile group with all piles the same size, the moment of inertia in the transverse direction, I_x, is

$$I_x = \sum d_{ix}^2 = (4)(5 \text{ ft})^2 + (4)(15 \text{ ft})^2 = 1000 \text{ ft}^2$$

The moment of inertia of the pile group in the longitudinal direction, I_y, is

$$I_y = \sum d_{iy}^2 = 8d_y^2$$

d_y is the distance from the centerline of the pile group to a transverse row of piles or half the longitudinal pile spacing.

The factored moment on the pile group due to the transverse force, M_x, is

$$M_x = F_x H = (50 \text{ kips})(20 \text{ ft}) = 1000 \text{ ft-kips}$$

The factored moment on the pile group due to the longitudinal force, M_y, is

$$M_y = F_y H = (30 \text{ kips})(20 \text{ ft}) = 600 \text{ ft-kips}$$

The maximum pile axial load, P, on one of the corner piles is

$$P = \frac{F_z}{\text{no. of piles}} + \frac{M_x d_{x,\max}}{I_x} + \frac{M_y d_{y,\max}}{I_y}$$

$$100 \text{ kips} = \frac{500 \text{ kips}}{8} + \frac{(1000 \text{ ft-kips})(15 \text{ ft})}{1000 \text{ ft}^2}$$

$$+ \frac{(600 \text{ ft-kips})d_y}{8d_y^2}$$

$$= 62.5 \text{ kips} + 15 \text{ kips} + \frac{75 \text{ ft-kips}}{d_y}$$

$$d_y = \frac{75 \text{ ft-kips}}{100 \text{ kips} - 62.5 \text{ kips} - 15 \text{ kips}}$$

$$= 3.33 \text{ ft}$$

The minimum required transverse pile spacing, s_y, is

$$s_y = 2d_y = (2)(3.33 \text{ ft})$$

$$= 6.67 \text{ ft} \quad (6 \text{ ft } 8 \text{ in}) \quad \text{[controls]}$$

From AASHTO Sec. 10.7.3.9, the center-to-center pile spacing should be at least 2.5 pile diameters.

$$s_{\min} = 2.5 D_{\text{pile}} = (2.5)(2 \text{ ft})$$

$$= 5 \text{ ft } 0 \text{ in} \quad \text{[does not control]}$$

The answer is (C).

83. The area, A, of the footing is

$$A = L_x L_y = (6 \text{ ft})(8 \text{ ft}) = 48 \text{ ft}^2$$

The section moduli in the x- and y-directions are

$$S_x = \frac{L_y L_x^2}{6} = \frac{(8 \text{ ft})(6 \text{ ft})^2}{6} = 48 \text{ ft}^3$$

$$S_y = \frac{L_x L_y^2}{6} = \frac{(6 \text{ ft})(8 \text{ ft})^2}{6} = 64 \text{ ft}^3$$

Check that the full-footing area is effective in resisting the axial load and moments.

The minimum bearing pressure, q_{min}, at the corner opposite from point A is

$$q_{min} = \frac{P}{A} - \frac{M_x}{S_x} - \frac{M_y}{S_y}$$
$$= \frac{200 \text{ kips}}{48 \text{ ft}^2} - \frac{40 \text{ ft-kips}}{48 \text{ ft}^3} - \frac{100 \text{ ft-kips}}{64 \text{ ft}^3}$$
$$= 1.77 \text{ kips/ft}^2 \geq 0 \text{ kips/ft}^2 \quad [\text{OK}]$$

The maximum bearing pressure, q_{max}, in tons is

$$q_{max} = \frac{P}{A} + \frac{M_x}{S_x} + \frac{M_y}{S_y}$$
$$= \frac{\dfrac{200 \text{ kips}}{48 \text{ ft}^2} + \dfrac{40 \text{ ft-kips}}{48 \text{ ft}^3} + \dfrac{100 \text{ ft-kips}}{64 \text{ ft}^3}}{2 \dfrac{\text{tons}}{\text{kip}}}$$
$$= 3.3 \text{ tons/ft}^2$$

The answer is (A).

84. The requirements for structural observation are given in IBC Sec. 1704.6.

From IBC Sec. 1704.6.1(1), structural observations are required for structures assigned to seismic design category D and risk category III or IV. Therefore, option A is incorrect.

From IBC Sec. 1704.6.1(3), structural observations are required for structures that are assigned to seismic design category E, are classified as risk category II, and are greater than two stories above grade plane. Therefore, option B is incorrect.

From IBC Sec. 1704.6.2(2), structural observations are required for structures where the ASD wind velocity is greater than 110 mph, and the building height is greater than 75 ft. Therefore, option D is incorrect.

From IBC Sec. 1704.6.1(2), structural observations are required for structures that are assigned to seismic design category D and have a building height greater than 75 ft above the base. For a building height of 60 ft, structural observations are not required.

The answer is (C).

Solutions
Lateral Forces Component: Buildings Depth Module Exam

85. (a) Determine the design spectral acceleration parameters per ASCE/SEI7 Sec. 11.4. For soil site class C with $S_S \geq 1.25$ and $S_1 \geq 0.5$, ASCE/SEI7 Table 11.4-1 and Table 11.4-2 give the short-period site coefficient, F_a, and the long-period site coefficient, F_v, as

$$F_a = 1.0$$
$$F_v = 1.3$$

From ASCE/SEI7 Eq. 11.4-1 and Eq. 11.4-2,

$$S_{MS} = F_a S_S = (1.0)(1.4) = 1.4$$
$$S_{M1} = F_v S_1 = (1.3)(0.53) = 0.689$$

From ASCE/SEI7 Eq. 11.4-3 and Eq. 11.4-4,

$$S_{DS} = \tfrac{2}{3} S_{MS} = \left(\tfrac{2}{3}\right)(1.4) = 0.933$$
$$S_{D1} = \tfrac{2}{3} S_{M1} = \left(\tfrac{2}{3}\right)(0.689) = 0.459$$

For a single-story building, the lateral seismic force at the roof level is equal to the seismic weight multiplied by the seismic response coefficient. The seismic weight, W, tributary to the roof is given as 600 kips.

Find the seismic response coefficient from ASCE/SEI7 Sec. 12.8. The period of the building, T_a, is given by ASCE/SEI7 Eq. 12.8-7. The approximate period parameters, C_t and x, are given in ASCE/SEI7 Table 12.8-2 under the category "for all other structural systems."

$$T_a = C_t h_n^x = (0.02)(18 \text{ ft})^{0.75} = 0.175 \text{ sec}$$

From ASCE/SEI7 Table 12.2-1 case A.7, the response modification factor for a special reinforced masonry shear bearing wall system is $R = 5$.

Find the seismic response coefficient, C_s, using ASCE/SEI7 Eq. 12.8-2.

$$C_s = \frac{S_{DS}}{\dfrac{R}{I_e}} = \frac{0.933}{\dfrac{5}{1}} = 0.187$$

When $T < T_L$, C_s need not exceed the value calculated using ASCE/SEI7 Eq. 12.8-3.

$$C_s = \frac{S_{D1}}{T\left(\dfrac{R}{I_e}\right)} = \frac{0.459}{(0.175)\left(\dfrac{5}{1}\right)} = 0.525$$

$0.525 > 0.187$, so 0.187 controls.

Per ASCE/SEI7 Eq. 12.8-5, C_s shall not be less than

$$C_s = 0.044 S_{DS} I_e = (0.044)(0.933)(1.0)$$
$$= 0.041 \geq 0.01 \quad [\text{OK}]$$

$0.041 < 0.187$; therefore, the seismic response coefficient is

$$C_s = 0.187$$

The problem specifies using the ASD load factor of $0.7E$. The lateral force at the roof level, F_{roof}, is

$$F_{\text{roof}} = 0.7 C_s W = (0.7)(0.187)(600 \text{ kips}) = 78.5 \text{ kips}$$

For a building with a flexible diaphragm, the lateral force is distributed by the tributary area. For seismic loads in the east-west direction, shear walls C and D will each resist half of the lateral seismic force.

$$V = \frac{F_{\text{roof}}}{2} = \frac{78.5 \text{ kips}}{2} = 39.3 \text{ kips}$$

For ASD analysis of special reinforced masonry shear walls, the in-plane shear force must be multiplied by 1.5 per TMS 402 Sec. 7.3.2.6.1.2.

$$V_{\text{ASD}} = 1.5 V = (1.5)(39.3 \text{ kips}) = \boxed{58.9 \text{ kips}}$$

85. (b) The masonry shear wall is evaluated per TMS 402 Sec. 8.3.5. Using the value for V_{ASD} calculated in part 85(a), per TMS 402 Eq. 8-24, the shear stress, f_v, in the wall is

$$f_v = \frac{V_{\text{ASD}}}{A_{nv}} = \frac{(58.9 \text{ kips})\left(1000 \dfrac{\text{lbf}}{\text{kip}}\right)}{\left(52.4 \dfrac{\text{in}^2}{\text{ft}}\right)(19 \text{ ft})}$$

$$= 59.2 \text{ lbf/in}^2$$

h is the height of the wall. The moment, M, at the base of the wall is then

$$M = Vh = (39.3 \text{ kips})(18 \text{ ft}) = 707 \text{ ft-kips}$$

(The moment, M, could be reduced by including the restorative moment due to the self-weight of the wall with a load factor of 0.6. However, the axial load is to be neglected, so the solution ignores the self-weight of the wall. The 1.5 multiplier of TMS 402 Sec. 7.3.2.6.1.2 does not need to be applied to the overturning moment.)

The ratio M/Vd is

$$\frac{M}{Vd} = \frac{707 \text{ ft-kips}}{(39.3 \text{ kips})(19 \text{ ft})} = 0.947 < 1.0$$

The allowable shear stress resisted by the masonry for special reinforced masonry shear walls is determined using TMS 402 Eq. 8-28. Conservatively disregard the axial load in the masonry.

$$F_{vm} = \frac{1}{4}\left(\left(4.0 - 1.75\left(\frac{M}{Vd}\right)\right)\sqrt{f_m'} \right) + 0.25\left(\frac{P}{A_n}\right)$$

$$= \left(\frac{1}{4}\right)\left((4.0 - (1.75)(0.947))\sqrt{1500 \frac{\text{lbf}}{\text{in}^2}} \right) + 0 \frac{\text{lbf}}{\text{in}^2}$$

$$= 22.7 \text{ lbf/in}^2$$

Since $F_{vm} < f_v$, steel reinforcement is required to resist the shear stress not resisted by the masonry. Therefore, from TMS 402 Eq. 8-25 and Eq. 8-30,

$$F_{vs} \geq \left(f_v - F_{vm}\right)\gamma_g$$

$\gamma_g = 0.75$ for partially grouted shear walls

$$\geq \left(59.2 \frac{\text{lbf}}{\text{in}^2} - 22.7 \frac{\text{lbf}}{\text{in}^2}\right)0.75$$

$$\geq 27.4 \text{ lbf/in}^2$$

The required spacing of no. 5 horizontal shear reinforcement (area of 0.31 in^2 per ACI 318 App. A) is calculated by rearranging TMS 402 Eq. 8-30. Per TMS 402

Sec. 8.3.3.1, the allowable steel stress, F_s, for grade 60 reinforcement is 32,000 lbf/in^2.

$$s = 0.5\left(\frac{A_v F_s d}{A_n(f_v - F_{vm})}\right)$$

$$= (0.5)\left(\frac{(0.31 \text{ in}^2)\left(32,000 \dfrac{\text{lbf}}{\text{in}^2}\right)(19 \text{ ft})\left(12 \dfrac{\text{in}}{\text{ft}}\right)}{\left(52.4 \dfrac{\text{in}^2}{\text{ft}}\right)(19 \text{ ft})\left(27.4 \dfrac{\text{lbf}}{\text{in}^2}\right)}\right)$$

$$= 31.1 \text{ in} \quad [\text{use 32 in}]$$

Check the requirements of TMS 402 Eq. 8-27. For $M/Vd \approx 1.0$

$$F_v \leq \left(2\sqrt{f_m'}\right)\gamma_g$$

$$59.2 \frac{\text{lbf}}{\text{in}^2} \leq \left(2\sqrt{1500 \frac{\text{lbf}}{\text{in}^2}}\right)0.75$$

$$\leq 58.1 \text{ lbf/in}^2$$

The requirement of Eq. 8-27 (b) is not quite met using the simplification of $M/Vd \approx 1.0$. By inspection, by using the actual value of $M/Vd \approx 0.947$ and linearly interpolating between Eq. 8-26 and Eq. 8-27 as permitted by Sec. 8.3.5.1.2, the requirements of this section are satisfied.

The minimum spacing must be in accordance with TMS Sec. 8.3.5.2.1, which states that the spacing may not exceed the lesser of

$$s_{\max} = \begin{cases} \dfrac{d}{2} = \dfrac{(19 \text{ ft})\left(12 \dfrac{\text{in}}{\text{ft}}\right)}{2} = 114 \text{ ft} > 32 \text{ in} \quad [\text{OK}] \\ 48 \text{ in} > 32 \text{ in} \quad [\text{OK}] \end{cases}$$

$\boxed{\text{Space the horizontal bars at 32 in o.c.}}$

85. (c) The wall is subjected to out-of-plane moments due to the eccentricity of the dead load and a uniformly distributed seismic load.

Find the design loads at the mid-height of the wall. Considering the maximum moment due to the out-of-plane earthquake load and half of the moment due to the eccentric dead load, the mid-height of the wall approximately corresponds to the location of maximum moment.

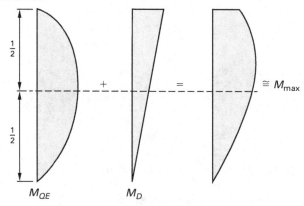

Use the short-period design spectral response acceleration, S_{DS}, calculated in part 85(a) to calculate the out-of-plane seismic load, q_{Q_E}. The weight of the wall, w_m, is given as 74 lbf/in². From ASCE/SEI7 Sec. 12.11.1, the out-of-plane seismic load is

$$q_{Q_E} = 0.4 S_{DS} I_e w_m$$
$$= (0.4)(0.933)(1.0)\left(74\ \frac{\text{lbf}}{\text{ft}^2}\right)$$
$$= 27.6\ \text{lbf/ft}^2 > (10\%) w_m$$

h is the height of the wall. The maximum moment per foot of wall due to seismic forces, M_{Q_E}, at the mid-height of the wall is

$$M_{Q_E} = \frac{q_{Q_E} h^2}{8} = \frac{\left(27.6\ \dfrac{\text{lbf}}{\text{ft}^2}\right)(18\ \text{ft})^2}{8} = 1118\ \text{ft-lbf/ft}$$

From problem illustration II, the eccentricity of the gravity load is

$$e = 3.5\ \text{in} + \frac{9.625\ \text{in}}{2} = 8.3125\ \text{in}$$

B is the width of the building. The moment per foot at the mid-height of the wall due to the eccentric dead load, M_D, is equal to half the dead load reaction on the wall times the eccentricity.

$$M_D = \frac{1}{2} q_{\text{dead}} \left(\frac{B}{2}\right) e = \left(\frac{1}{2}\right)\left(25\ \frac{\text{lbf}}{\text{ft}^2}\right)\left(\frac{50\ \text{ft}}{2}\right)\left(\frac{8.3125\ \text{in}}{12\ \dfrac{\text{in}}{\text{ft}}}\right)$$
$$= 216\ \text{ft-lbf/ft}$$

Vertical and horizontal seismic load effects are covered in ASCE/SEI7 Sec. 12.4. From ASCE/SEI7 Sec. 12.4.2.3,

allowable stress design load combination $D + 0.7E$ can be rewritten as

$$(1.0 + 0.14 S_{DS}) D + 0.7 \rho Q_E$$

The design out-of-plane moment, M, for shear wall A is

$$M = (1.0 + 0.14 S_{DS}) M_D + 0.7 \rho M_{Q_E}$$
$$= \left(1.0 + (0.14)(0.933)\right)\left(216\ \frac{\text{ft-lbf}}{\text{ft}}\right)$$
$$+ (0.7)(1.0)\left(1118\ \frac{\text{ft-lbf}}{\text{ft}}\right)$$
$$= \boxed{1030\ \text{ft-lbf/ft}}$$

The axial force, P_D, at the mid-height of the wall is equal to the roof dead load reaction plus the self-weight of the top half of the wall.

$$P_D = q_{\text{dead}} \left(\frac{B}{2}\right) + w_m \left(\frac{h}{2}\right)$$
$$= \left(25\ \frac{\text{lbf}}{\text{ft}^2}\right)\left(\frac{50\ \text{ft}}{2}\right) + \left(74\ \frac{\text{lbf}}{\text{ft}^2}\right)\left(\frac{18\ \text{ft}}{2}\right)$$
$$= 1290\ \text{lbf/ft}$$

Using the load combination from ASCE/SEI7 Sec. 12.4.2.3, the design axial load, P, at the mid-height of the wall is

$$P_{\text{mid-height}} = (1.0 + 0.14 S_{DS}) P_D$$
$$= \left(1.0 + (0.14)(0.933)\right)\left(1290\ \frac{\text{lbf}}{\text{ft}}\right)$$
$$= \boxed{1460\ \text{lbf/ft}}$$

85. (d) The moment and axial loads at the mid-height of the wall due to load combination $D + 0.7E$ were calculated in part 85(c) as

$$M_{\text{mid-height}} = 1030\ \text{ft-lbf/ft}$$
$$P_{\text{mid-height}} = 1460\ \text{lbf/ft}$$

The stress due to the axial component of the load, f_a, is

$$f_a = \frac{P_{\text{mid-height}}}{A_n} = \frac{1460\ \dfrac{\text{lbf}}{\text{ft}}}{52.4\ \dfrac{\text{in}^2}{\text{ft}}} = 27.9\ \text{lbf/in}^2$$

Check the compressive force due to only the axial load against the allowable loads of TMS 402 Sec. 8.3.4.2.1.

From ACI 318 App. A, the nominal area of no. 5 bars is 0.31 in^2. The area, A_{st}, of steel in a one-foot strip of wall is

$$A_{st} = (0.31 \text{ in}^2)\left(\dfrac{12 \frac{\text{in}}{\text{ft}}}{32 \text{ in}}\right) = 0.116 \text{ in}^2/\text{ft}$$

The radius of gyration, r, is

$$r = \sqrt{\dfrac{I_n}{A_n}} = \sqrt{\dfrac{624.6 \frac{\text{in}^4}{\text{ft}}}{52.4 \frac{\text{in}^2}{\text{ft}}}} = 3.45 \text{ in}$$

The h/r ratio is

$$\dfrac{h}{r} = \dfrac{(18 \text{ ft})\left(12 \frac{\text{in}}{\text{ft}}\right)}{3.45 \text{ in}} = 62.6$$

Since $62.6 < 99$, the wall is not slender and TMS 402 Eq. 8-21 applies. Ignoring the contribution from the steel, the allowable axial compressive force, P_a, in the reinforced masonry is

$$P_a = (0.25 f'_m A_n)\left[1 - \left(\dfrac{h}{140r}\right)^2\right]$$

$$= \left[(0.25)\left(1500 \frac{\text{lbf}}{\text{in}^2}\right)\left(52.4 \frac{\text{in}^2}{\text{ft}}\right)\right]$$

$$\times \left[1 - \left(\dfrac{(18 \text{ ft})\left(12 \frac{\text{in}}{\text{ft}}\right)}{(140)(3.45 \text{ in})}\right)^2\right]$$

$$= 15,720 \text{ lbf/ft}$$

$P_a > P_{\text{mid-height}}$; therefore, the wall is adequate for axial loads alone at the mid-height of the wall.

Check the stresses in the masonry and steel due to the out-of-plane moment. Use the design assumptions of TMS 402 Sec. 8.3.2. Consider a 32 in width of wall, and assume that the neutral axis lies within the face shell of the masonry ($kd < 1.25$ in).

The vertical bar spacing, b, is

$$b = 32 \text{ in}$$

From problem illustration II, the depth of the reinforcement bar, d, at the center of the wall thickness is

$$d = \dfrac{9.625 \text{ in}}{2} = 4.81 \text{ in}$$

The moment, M, is

$$M = M_{\text{mid-height}} b = \left(1030 \frac{\text{ft-lbf}}{\text{ft}}\right)(32 \text{ in})$$

$$= 33,000 \text{ in-lbf}$$

The area of a no. 5 reinforcing bar, A_{st}, is

$$A_{st} = 0.31 \text{ in}^2$$

The reinforcement ratio, ρ, is

$$\rho = \dfrac{A_{st}}{bd} = \dfrac{0.31 \text{ in}^2}{(32 \text{ in})(4.81 \text{ in})} = 0.00201$$

Per TMS 402 Sec. 4.2.2.1, the modulus of elasticity of steel reinforcement, E_s, is

$$E_s = 29,000,000 \text{ lbf/in}^2$$

Per TMS 402 Sec. 4.2.2.2.1, the modulus of elasticity for concrete masonry, E_m, is

$$E_m = 900 f'_m = (900)\left(1500 \frac{\text{lbf}}{\text{in}^2}\right) = 1,350,000 \text{ lbf/in}^2$$

The modular ratio, n, is

$$n = \dfrac{E_s}{E_m} = \dfrac{29,000,000 \frac{\text{lbf}}{\text{in}^2}}{1,350,000 \frac{\text{lbf}}{\text{in}^2}} = 21.48$$

The product of the modular ratio and the reinforcement ratio is

$$n\rho = (21.48)(0.00201) = 0.0432$$

The neutral axis depth factor, k, is

$$k = \sqrt{(n\rho)^2 + 2n\rho} - n\rho$$

$$= \sqrt{(0.0432)^2 + (2)(0.0432)} - 0.0432$$

$$= 0.254$$

Check that the neutral axis is within the face shell.

$$kd = (0.254)(4.81 \text{ in}) = 1.22 \text{ in} < 1.25 \text{ in} \quad [\text{OK}]$$

The lever-arm factor, j, is

$$j = 1 - \frac{k}{3} = 1 - \frac{0.254}{3} = 0.915$$

The stress in the steel reinforcement, f_s, due to the applied moment is

$$f_s = \frac{M}{A_{st}jd} = \frac{33{,}000 \text{ in-lbf}}{(0.31 \text{ in}^2)(0.915)(4.81 \text{ in})} = 24{,}200 \text{ lbf/in}^2$$

Per TMS 402 Sec. 8.3.3.1, the allowable steel tensile stress, F_s, for grade 60 reinforcement is

$$F_s = 32{,}000 \text{ lbf/in}^2 > 24{,}200 \text{ lbf/in}^2 \quad \text{[OK]}$$

The maximum stress in the masonry due to the applied moment, f_b, is

$$\begin{aligned} f_b &= \frac{2M}{jkbd^2} \\ &= \frac{(2)(33{,}000 \text{ in-lbf})}{(0.915)(0.254)(32 \text{ in})(4.81 \text{ in})^2} \\ &= 383 \text{ lbf/in}^2 \end{aligned}$$

To be conservative, directly combine the compressive stress due to flexure and direct axial loads. The total masonry stress is

$$f_b + f_a = 383 \ \frac{\text{lbf}}{\text{in}^2} + 28 \ \frac{\text{lbf}}{\text{in}^2} = 411 \text{ lbf/in}^2$$

Per TMS 402 Sec. 8.3.4.2.2, the allowable masonry compressive stress, F_b, is

$$\begin{aligned} F_b &= 0.45 f'_m = (0.45)\left(1500 \ \frac{\text{lbf}}{\text{in}^2}\right) \\ &= 675 \text{ lbf/in}^2 > 411 \text{ lbf/in}^2 \quad \text{[OK]} \end{aligned}$$

TMS 402 Sec. 8.3.4.2.2 requires that the compressive stress in the masonry due to the axial load component, f_a, not exceed the allowable stress, F_a, of Sec. 8.2.4.1. By inspection, the compressive stress due to the axial load component, $f_a = 27.9 \text{ lbf/in}^2$, does not exceed the allowable stress, F_a.

> The wall is adequate for combined compression and flexure.

86. (a) Wind loads are determined using the envelope procedure in accordance with ASCE/SEI7 Chap. 28, Part 1. Find the velocity pressure, q_z, using ASCE/SEI7 Eq. 28.3-1. (ASCE/SEI7 uses V for wind speed, whereas this solution uses v. This equation is not dimensionally consistent.)

$$q_{z,\text{lbf/ft}^2} = 0.00256 K_z K_{zt} K_d \text{v}^2$$

The roof height, z, is given as 13 ft. From ASCE/SEI7 Table 28.3-1, the velocity pressure exposure coefficient, K_z, for exposure C and $z < 15$ ft is

$$K_z = 0.85$$

The building sits at the top of an escarpment, so the topographic factor, K_{zt}, must be calculated in accordance with ASCE/SEI7 Sec. 26.8 and Fig. 26.8-1.

Use problem illustration II to find the values of variables needed to calculate the topographic factor.

The distance upwind of the crest to where the difference in ground elevation is half the height of the escarpment, L_h, is equal to half the escarpment length.

$$L_h = \frac{100 \text{ ft}}{2} = 50 \text{ ft}$$

The height, H, of the escarpment is 20 ft. So, the first multiplier from ASCE/SEI7 Fig. 26.8-1 is

$$\frac{H}{L_h} = \frac{20 \text{ ft}}{50 \text{ ft}} = 0.40 \quad (K_1 = 0.34)$$

The distance from the crest to the building site, x, is 25 ft. So, the second multiplier from ASCE/SEI7 Fig. 26.8-1 is

$$\frac{x}{L_h} = \frac{25 \text{ ft}}{50 \text{ ft}} = 0.50 \quad (K_2 = 0.88)$$

The height of the building above local ground level, z, is 13 ft. So, the third multiplier from ASCE/SEI7 Fig. 26.8-1 is

$$\frac{z}{L_h} = \frac{13 \text{ ft}}{50 \text{ ft}} = 0.26 \quad (K_3 \approx 0.53)$$

From ASCE/SEI7 Eq. 26.8-1, K_{zt} is

$$\begin{aligned} K_{zt} &= (1 + K_1 K_2 K_3)^2 \\ &= (1 + (0.34)(0.88)(0.53))^2 \\ &= 1.34 \end{aligned}$$

From ASCE/SEI7 Table 26.6-1, the wind directionality factor for the wind load on the main wind force-resisting system (MWFRS) is

$$K_d = 0.85$$

Therefore, the velocity pressure at the roof height, $q_{z,\text{lbf/ft}^2}$, is

$$q_{z,\text{lbf/ft}^2} = q_h = 0.00256 K_z K_{zt} K_d \text{v}^2$$
$$= (0.00256)(0.85)(1.34)(0.85)\left(140\ \frac{\text{mi}}{\text{hr}}\right)^2$$
$$= 48.6\ \text{lbf/ft}^2$$

The design pressures for the MWFRS of a low-rise building are determined per ASCE/SEI7 Sec. 28.4-1.

The external pressure coefficient, GC_{pf}, is determined from ASCE/SEI7 Fig. 28.4-1. For load case A zone 1E and load case B zone 5E (corner zone of windward wall) and a roof angle of $\theta = 0°$,

$$GC_{pf} = +0.61$$

From ASCE/SEI7 Table 26.11-1, the internal pressure coefficient, GC_{pi}, for an enclosed building is

$$GC_{pi} = \pm 0.18$$

From ASCE/SEI7 Eq. 28.4-1, the worst case pressure, p, on a corner zone of the south wall is

$$p = q_h(GC_{pf} - GC_{pi}) = \left(48.6\ \frac{\text{lbf}}{\text{ft}^2}\right)(0.61 - (-0.18))$$
$$= \boxed{38\ \text{lbf/ft}^2}$$

(For ASD, the above strength design pressure would get multiplied by 0.6.)

86. (b) The double top plate will act as a collector for wind loads in the north-south direction and as a diaphragm chord for wind loads in the east-west direction. Check the wind loads in both directions to determine which wind load controls.

B is the width of the building perpendicular to the wind load. For the north-south wind load case (i.e., the top plate acting as a collector element), determine the reaction, R, at the east end of the diaphragm.

$$R = w_{\text{N-S}}\left(\frac{B}{2}\right) = \left(225\ \frac{\text{lbf}}{\text{ft}}\right)\left(\frac{50\ \text{ft}}{2}\right) = 5625\ \text{lbf}$$

The unit shear of the roof diaphragm reaction, v_{dia}, is equal to the reaction divided by the length, L, of the diaphragm parallel to the wind load.

$$v_{\text{dia}} = \frac{R}{L} = \frac{5625\ \text{lbf}}{25\ \text{ft}} = 225\ \text{lbf/ft}$$

The maximum collector force in the top plate, T, occurs at the north end of the collector and is equal to the length of the collector element, L_c, multiplied by the diaphragm unit shear.

$$T = L_c v_{\text{dia}} = (10\ \text{ft})\left(225\ \frac{\text{lbf}}{\text{ft}}\right) = 2250\ \text{lbf}\quad[\text{controls}]$$

For the east-west wind load case (i.e., the top plate acting as a diaphragm chord), the maximum moment, M, in a simply supported flexible diaphragm is

$$M = \frac{w_{\text{E-W}}L^2}{8} = \frac{\left(400\ \frac{\text{lbf}}{\text{ft}}\right)(25\ \text{ft})^2}{8} = 31{,}250\ \text{ft-lbf}$$

The maximum tensile force in the diaphragm chord, T, is equal to the moment divided by the spacing of the top plates.

$$T = \frac{M}{B} = \frac{31{,}250\ \text{ft-lbf}}{50\ \text{ft}} = 625\ \text{lbf}$$

2250 lbf > 625 lbf; therefore, the wind load in the north-south direction controls.

The splice of the double top plate should be designed for a maximum tension of $\boxed{2250\ \text{lbf.}}$

86. (c) From problem part 86(b), use a top plate tensile force of 2250 lbf. The tensile force will load the nails in shear. The problem illustration indicates that the given wind loads are at service-level, so a load factor of 0.6 per IBC 1605.3.1 is not required.

The adjustment factors for dowel-type fasteners are given in NDS Table 10.3.1. C_{eg}, C_{di}, and C_{tn} are not applicable. As given in the problem statement,

$$C_M = C_t = 1.0$$

For fasteners where $D < \frac{1}{4}$ in, per NDS Sec. 10.3.6 and Sec. 11.5.1,

$$C_g = C_\Delta = 1.0$$

For wind loads per NDS Table 2.3.2,

$$C_D = 1.6$$

Reference lateral design values for nailed sawn lumber connections loaded in single shear are given in NDS Table 11N. For eastern softwoods, the specific gravity, SG, is 0.36. (NDS and SDPWS use G for specific gravity whereas this solution uses SG.) The 2 in × 6 in top plate has a thickness, t_s, of 1.5 in. From Table 11N, the lateral design value for 8d box nails is

$$Z = 54\ \text{lbf}$$

According to NDS Table 11N, ftn. 4, the nail length is insufficient to provide $10D$ penetration, so the lateral design value must be adjusted per NDS Table 11N, ftn. 3. The length, L_{nail}, of an 8d box nail is given in NDS App. L, Table L4, as 2.5 in. Therefore, the penetration depth, p, is

$$p = L_{\text{nail}} - t_s = 2.5 \text{ in} - 1.5 \text{ in} = 1 \text{ in}$$

From NDS Table 11N, ftn. 3, Z must be multiplied by $p/(10D)$.

$$\frac{p}{10D} = \frac{1 \text{ in}}{(10)(0.113 \text{ in})} = 0.885$$

The adjusted lateral design value, Z', is then

$$Z' = ZC_D\left(\frac{p}{10D}\right) = (54 \text{ lbf})(1.6)(0.885) = 76 \text{ lbf}$$

The required number of nails, n, is equal to the maximum top plate tensile force, T, divided by the adjusted lateral design value of a single nail.

$$n = \frac{T}{Z'} = \frac{2250 \text{ lbf}}{76 \text{ lbf}} = \boxed{29.6 \quad [\text{use 30 nails}]}$$

86. (d) The shear in wall A, v_A, due to the north-south wind load is the diaphragm reaction, R, from part 86 (b), divided by the shear wall length, b.

$$v_A = \frac{R}{b} = \frac{5625 \text{ lbf}}{15 \text{ ft}} = 375 \text{ lbf/ft}$$

The nominal unit shear capacities for wood-framed shear walls with wood-based panels are given in SDPWS Table 4.3A. From SDPWS Table 4.3A, ftn. 3, the tabulated values for species other than Douglas fir-larch or southern pine must be adjusted for specific gravity.

From NDS Table 11.3.2A, for eastern softwoods, the specific gravity is

$$SG = 0.36$$

The specific gravity adjustment factor, C_{SG}, is calculated using the equation in SDPWS Table 4.3A, ftn. 3.

$$C_{SG} = 1 - (0.5 - SG) = 1 - (0.5 - 0.36) = 0.86$$

From SDPWS Sec. 4.3.3 for ASD, the nominal unit shear capacities must be divided by 2.0. To find the required shear capacity of the wall for wind loads, multiply the shear in wall A by 2.0, and then divide by the specific gravity adjustment factor.

$$v_w \geq \frac{2v_A}{C_{SG}} = \frac{(2)\left(375 \dfrac{\text{lbf}}{\text{ft}}\right)}{0.86} = 872 \text{ lbf/ft}$$

Select a nailing pattern from SDPWS Table 4.3A where the nominal unit shear wind capacity, v_w, exceeds 872 lbf/ft. The required nail spacing along intermediate framing members is given in SDPWS Sec. 4.3.7.1.

Acceptable sheathing and nailing patterns are listed below. Note that only one design is required.

- $^{19}/_{32}$ in wood structural panel sheathing (4 ft × 8 ft) with 10d common or galvanized box nails. Nails should be spaced at 6 in o.c. at panel edges and 12 in o.c. along intermediate framing members. Provide blocking or framing members at edges of all panels.

$$v_w = 950 \frac{\text{lbf}}{\text{ft}} > 872 \frac{\text{lbf}}{\text{ft}}$$

- $^3/_8$ in wood structural panel sheathing (4 ft × 8 ft) with 8d common or galvanized box nails. Nails should be spaced at 4 in o.c. at panel edges and 6 in o.c. along intermediate framing members. Provide blocking or framing members at edges of all panels.

$$v_w = 895 \frac{\text{lbf}}{\text{ft}} > 872 \frac{\text{lbf}}{\text{ft}}$$

- $^5/_{16}$ in wood structural panel sheathing (4 ft × 8 ft) with 6d common or galvanized box nails. Nails should be spaced at 3 in o.c. at panel edges and 6 in o.c. along intermediate framing members. Provide blocking or framing members at edges of all panels.

$$v_w = 980 \frac{\text{lbf}}{\text{ft}} > 872 \frac{\text{lbf}}{\text{ft}}$$

Other shear wall designs are also acceptable, provided that the tabulated allowable shear value does not exceed the adjusted calculated demand.

86. (e) A sketch of the connection detail between the roof framing and shear wall A is shown.

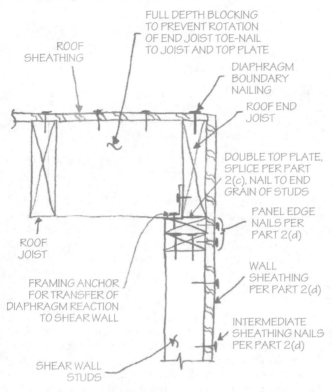

87. (a) Determine the distribution of forces to the lateral-load resisting elements of the rigid diaphragm building. Assume an origin at gridline A-1. The relative rigidities and coordinates of the lateral load-resisting elements are

- moment frame on gridline 1, $R_1 = 1.0$; $x_1 = 0$ ft
- moment frame on gridline 4, $R_4 = 1.0$; $x_4 = 62$ ft
- shear wall C, $R_C = 5.0$; $y_C = 50$ ft
- shear wall D, $R_D = 5.0$; $y_D = 75$ ft

From problem illustration I, the center of mass, (x_{CM}, y_{CM}), is located at

$$x_{CM} = 31 \text{ ft}$$
$$y_{CM} = 50 \text{ ft}$$

The center of rigidity, (x_{CR}, y_{CR}), is located at

$$x_{CR} = \frac{x_1 + x_4}{2} = \frac{0 \text{ ft} + 62 \text{ ft}}{2} = 31 \text{ ft}$$
$$y_{CR} = \frac{y_C + y_D}{2} = \frac{50 \text{ ft} + 75 \text{ ft}}{2} = 62.5 \text{ ft}$$

The polar moment of inertia in the x-direction (east-west), J_x, is

$$J_x = R_1(x_1 - x_{CR})^2 + R_4(x_4 - x_{CR})^2$$
$$= (1.0)(0 \text{ ft} - 31 \text{ ft})^2 + (1.0)(62 \text{ ft} - 31 \text{ ft})^2$$
$$= 1922 \text{ ft}^2$$

The polar moment of inertia in the y-direction (north-south), J_y, is

$$J_y = R_C(y_C - y_{CR})^2 + R_D(y_D - y_{CR})^2$$
$$= (5.0)(50 \text{ ft} - 62.5 \text{ ft})^2 + (5.0)(75 \text{ ft} - 62.5 \text{ ft})^2$$
$$= 1563 \text{ ft}^2$$

The total polar moment of inertia, J, is

$$J = J_x + J_y = 1922 \text{ ft}^2 + 1563 \text{ ft}^2 = 3485 \text{ ft}^2$$

The total torsional moment, M_T, is the sum of the moment due to 1) the eccentricity between the center of rigidity and the center of mass and 2) the accidental torsion per ASCE/SEI7 Sec. 12.8.4.2.

$$M_T = F_y(y_{CR} - y_{CM}) + 0.05 F_y L_y$$
$$= (500 \text{ kips})(62.5 \text{ ft} - 50 \text{ ft})$$
$$\quad + (0.05)(500 \text{ kips})(100 \text{ ft})$$
$$= 8750 \text{ ft-kips}$$

The shear force, V_u, in wall C is

$$V_u = \frac{F_y R_C}{R_C + R_D} + \frac{M_T R_C(y_{CR} - y_C)}{J}$$
$$= \frac{(500 \text{ kips})(5.0)}{5.0 + 5.0}$$
$$\quad + \frac{(8750 \text{ ft-kips})(5.0)(62.5 \text{ ft} - 50 \text{ ft})}{3485 \text{ ft}^2}$$
$$= \boxed{407 \text{ kips}}$$

87. (b) The requirements for special structural concrete walls are covered in ACI 318 Sec. 18.10.2.

From part 87(a), the shear force in wall C is $V_u = 407$ kips.

The gross concrete shear area, A_{cv}, of wall C is

$$A_{cv} = l_w h = (12 \text{ ft})(18 \text{ in})\left(12 \frac{\text{in}}{\text{ft}}\right) = 2592 \text{ in}^2$$

From ACI 318 Table 19.2.4.2, $\lambda = 1.0$ for normal weight concrete.

From ACI 318 Sec. 18.10.2.1,

$$A_{cv}\lambda\sqrt{f_c'} = \frac{(2592 \text{ in}^2)(1.0)\sqrt{5000 \frac{\text{lbf}}{\text{in}^2}}}{1000 \frac{\text{lbf}}{\text{kip}}} = 183 \text{ kips}$$

From part 87(a), V_u is 407 kips, which is greater than $A_{cv}\lambda\sqrt{f_c'}$. So, the reinforcement ratios, ρ_l and ρ_t, cannot be less than 0.0025 per ACI 318 Sec. 18.10.2.1.

From ACI 318 App. A, the area, A_b, of a no. 6 bar is 0.44 in^2. From ACI 318 Sec. 18.10.2.1, for two layers of horizontal no. 6 bars at 8 in o.c., the horizontal reinforcement, ρ_t, is

$$\rho_t = \frac{2A_b}{hs} = \frac{(2)(0.44 \text{ in}^2)}{(18 \text{ in})(8 \text{ in})} = \boxed{0.00611 > 0.0025 \quad [\text{OK}]}$$

From ACI 318 App. A, the area, A_b, of a no. 5 bar is 0.31 in^2. For two layers of vertical no. 5 bars at 12 in o.c., the vertical reinforcement, ρ_l, is

$$\rho_l = \frac{2A_b}{hs} = \frac{(2)(0.31 \text{ in}^2)}{(18 \text{ in})(12 \text{ in})} = \boxed{0.00287 > 0.0025 \quad [\text{OK}]}$$

From ACI 318 Sec. 18.10.4.1, the coefficient α_c

- is 3.0 for $h_w/l_w \leq 1.5$

- is 2.0 for $h_w/l_w \geq 2.0$

- varies linearly between 3.0 and 2.0 for $1.5 < h_w/l_w < 2.0$

The height, h_w, is given in the design criteria as 60 ft, and the wall length, l_w, is shown in problem illustration I as 12 ft.

$$\frac{h_w}{l_w} = \frac{60 \text{ ft}}{12 \text{ ft}} = 5.0 \geq 2.0 \quad (\alpha_c = 2.0)$$

The nominal strength of the wall, V_n, cannot exceed the value given by ACI 318 Eq. 18.10.4.1.

$$V_n = A_{cv}(\alpha_c\lambda\sqrt{f_c'} + \rho_t f_y)$$

$$= \frac{(2592 \text{ in}^2)\left(\begin{array}{c}(2.0)(1.0)\sqrt{5000 \frac{\text{lbf}}{\text{in}^2}} \\ +(0.00611)\left(60{,}000 \frac{\text{lbf}}{\text{in}^2}\right)\end{array}\right)}{1000 \frac{\text{lbf}}{\text{kip}}}$$

$$= 1320 \text{ kips}$$

From ACI 318 Sec. 18.10.4.4, calculate the maximum nominal shear strength for an individual vertical wall segment. (This equation is not dimensionally consistent.)

$$V_{n,\max} = 10A_{cw}\sqrt{f_c'} = \frac{(10)(2592 \text{ in}^2)\sqrt{5000 \frac{\text{lbf}}{\text{in}^2}}}{1000 \frac{\text{lbf}}{\text{kip}}}$$

$$= 1830 \text{ kips} > 1320 \text{ kips}$$

Therefore, $V_n = 1320$ kips.

Conservatively assume that the nominal shear strength of the member is less than the shear corresponding to the development of the nominal flexural strength. From ACI 318 Sec. 21.2.4.1, $\phi = 0.60$, so

$$\phi V_n = (0.60)(1320 \text{ kips}) = \boxed{790 \text{ kips} > 407 \text{ kips}}$$

The horizontal reinforcement of shear wall C is adequate.

87. (c) From ACI 318 Sec. 18.7.3.2, the flexural strengths of columns must satisfy ACI 318 Eq. 21-1.

$$\sum M_{nc} \geq \frac{6}{5}\sum M_{nb}$$

ACI 318 Sec. 18.7.3.2 defines M_{nc} as the nominal flexural strengths of the columns framing into the joint and M_{nb} as the nominal flexural strengths of the beams framing into the joint.

For a lateral load acting on the moment frame, the beam on one side of the joint will be in positive flexure, while the beam on the other side of the joint will be in negative flexure.

For negative flexure (top bars in tension), ACI 318 Sec. 18.7.3.2 requires that the slab reinforcement within an effective slab width as defined in ACI 318 Sec. 6.3.2 be considered when calculating the beam flexural strength. From ACI 318 Table 6.3.2.1, for beams with a slab on one side only, the effective overhanging flange width, $b_{f,\text{overhang}}$, is the smallest of one-twelfth the beam span, six times the slab thickness, or half the clear distance to the next web (which does not control by inspection).

$$b_{f,\text{overhang}} = \min \begin{cases} \left(\dfrac{1}{12}\right)(25\text{ ft}) = 2.08\text{ ft} \quad \text{[controls]} \\ (6)\left(\dfrac{8\text{ in}}{12\frac{\text{in}}{\text{ft}}}\right) = 4\text{ ft} \\ 1/2 \text{ clear distance to the next web} \end{cases}$$

From problem illustration II, the slab is reinforced with no. 6 bars at 9 in o.c. Therefore, there are three no. 6 slab bars in the 2.08 ft width of overhanging flange. For negative flexure, per ACI 318 App. A, the area, A_s, of four no. 9 beam top bars and three no. 6 slab bars is

$$A_s = (4)(1.0\text{ in}^2) + (3)(0.44\text{ in}^2) = 5.32\text{ in}^2$$

The depth of the compression block, a, is

$$a = \frac{A_s f_y}{0.85 f_c' b} = \frac{(5.32\text{ in}^2)\left(60{,}000\frac{\text{lbf}}{\text{in}^2}\right)}{(0.85)\left(5000\frac{\text{lbf}}{\text{in}^2}\right)(20\text{ in})} = 3.76\text{ in}$$

The nominal flexural strength, M_n^-, is

$$M_n^- = A_s f_y\left(d - \frac{a}{2}\right)$$
$$= \frac{(5.32\text{ in}^2)\left(60{,}000\frac{\text{lbf}}{\text{in}^2}\right)\left(21.5\text{ in} - \frac{3.76\text{ in}}{2}\right)}{\left(12\frac{\text{in}}{\text{ft}}\right)\left(1000\frac{\text{lbf}}{\text{kip}}\right)}$$
$$= 522\text{ ft-kips}$$

For positive flexure, the area, A_s, of three no. 9 beam bottom bars is

$$A_s = (3)(1.0\text{ in}^2) = 3\text{ in}^2$$

The compression block width, b, is

$$b = b_{\text{beam}} + b_{f,\text{overhang}} = 20\text{ in} + (2.08\text{ ft})\left(12\frac{\text{in}}{\text{ft}}\right)$$
$$= 45\text{ in}$$

(Using a width of $b = 20$ in is also acceptable and would not significantly alter the nominal positive flexural capacity.)

The depth of the compression block, a, is

$$a = \frac{A_s f_y}{0.85 f_c' b} = \frac{(3\text{ in}^2)\left(60{,}000\frac{\text{lbf}}{\text{in}^2}\right)}{(0.85)\left(5000\frac{\text{lbf}}{\text{in}^2}\right)(45\text{ in})} = 0.941\text{ in}$$

The nominal flexural strength, M_n^+, is

$$M_n^+ = A_s f_y\left(d - \frac{a}{2}\right)$$
$$= \frac{(3\text{ in}^2)\left(60{,}000\frac{\text{lbf}}{\text{in}^2}\right)\left(21.5\text{ in} - \frac{0.941\text{ in}}{2}\right)}{\left(12\frac{\text{in}}{\text{ft}}\right)\left(1000\frac{\text{lbf}}{\text{kip}}\right)}$$
$$= 315\text{ ft-kips}$$

The summation of the nominal flexural strengths of the beams framing into the joint is

$$\sum M_{nb} = M_n^- + M_n^+ = 522\text{ ft-kips} + 315\text{ ft-kips}$$
$$= 837\text{ ft-kips}$$

From ACI 318 Sec. 18.7.3.2, when calculating the nominal flexural strengths, a corresponding factored axial load must be considered. The nominal column moment capacities for the given factored axial loads are approximated from the interaction diagram given in problem illustration III.

For an axial load 300 kips above the floor level,

$$M_{nc,\text{above}} \approx 580\text{ ft-kips}$$

For an axial load of 350 kips below the floor level,

$$M_{nc,\text{below}} \approx 600\text{ ft-kips}$$

The summation of the nominal flexural strengths of the columns framing into the joint is

$$\sum M_{nc} = M_{nc,\text{above}} + M_{nc,\text{below}}$$
$$= 580\text{ ft-kips} + 600\text{ ft-kips}$$
$$= 1180\text{ ft-kips}$$

From ACI 318 Eq. 18.7.3.2,

$$\sum M_{nc} \geq \frac{6}{5}\sum M_{nb}$$
$$1180\text{ ft-kips} \geq \left(\frac{6}{5}\right)(837\text{ ft-kips})$$
$$\boxed{1180\text{ ft-kips} \geq 1000\text{ ft-kips}}$$

The requirements of ACI 318 Sec. 21.6.2.2 are met.

87. (d) From ACI 318 Sec. 18.8.2.2, beam longitudinal reinforcement that ends in a column must be extended to the far face of the confined column core and anchored in tension, according to ACI 318 Sec. 18.8.5.

Per ACI 318 App. A, the nominal diameter, d_b, of a no. 9 bar is 1.128 in. From ACI 318 Sec. 18.8.5.1, the development length for a hooked bar in tension, l_{dh}, is the largest of 6 in, $8d_b$, and ACI 318 Eq. 18.8.5.1.

$$l_{dh} = \max \begin{cases} 6 \text{ in} \\ 8d_b = (8)(1.128 \text{ in}) = 9.0 \text{ in} \\ \dfrac{f_y d_b}{65\sqrt{f_c'}} = \dfrac{\left(60{,}000 \, \dfrac{\text{lbf}}{\text{in}^2}\right)(1.128 \text{ in})}{(65)\sqrt{5000 \, \dfrac{\text{lbf}}{\text{in}^2}}} \\ = 14.7 \text{ in} \quad \text{[controls]} \end{cases}$$

Considering any reasonable column clear cover, this length is less than the dimension of the confined column core of a 24 in column. Therefore, the no. 9 beam longitudinal reinforcement can be fully developed in tension with a standard 90° hook at an end moment frame joint.

87. (e) A sketch of an elevation of the end moment frame joint at gridline A-1 is shown.

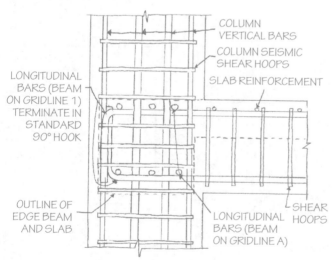

COLUMN VERTICAL BARS

COLUMN SEISMIC SHEAR HOOPS

SLAB REINFORCEMENT

LONGITUDINAL BARS (BEAM ON GRIDLINE 1) TERMINATE IN STANDARD 90° HOOK

LONGITUDINAL BARS (BEAM ON GRIDLINE A)

SHEAR HOOPS

OUTLINE OF EDGE BEAM AND SLAB

88. (a) Use ASCE/SEI7 Sec. 12.4.2.3, load combination 5.

$$(1.2 + 0.2S_{DS})D + \rho Q_E + 0.5L + 0.2S$$

In accordance with ASCE/SEI7 Sec. 12.4.2.3, note 1, the load factor on L is taken as 0.5 for live loads that do not exceed 100 lbf/ft².

With a redundancy factor of $\rho = 1.0$, the factored load at footing B-1 is

$$P_u = (1.2 + 0.2S_{DS})P_D + P_{Q_E} + 0.5P_L + 0.2P_S$$

P_D, P_{Q_E}, P_L, and P_S are the footing loads due to dead, horizontal seismic, live, and snow loads, respectively.

From problem illustration I, the column at gridline B-1 has a tributary area at each floor of

$$A = (16 \text{ ft})(16 \text{ ft}) = 256 \text{ ft}^2$$

The unfactored load at the footing due to the dead load, P_D, is

$$P_D = (\text{floors affected})A\,q_{\text{dead}} = \frac{(4)(256 \text{ ft}^2)\left(150 \, \dfrac{\text{lbf}}{\text{ft}^2}\right)}{1000 \, \dfrac{\text{lbf}}{\text{kip}}}$$

$$= 153.6 \text{ kips}$$

The unfactored load at the footing due to the snow load, P_S, is

$$P_S = (\text{floors affected})A\,q_{\text{snow}} = \frac{(1)(256 \text{ ft}^2)\left(30 \, \dfrac{\text{lbf}}{\text{ft}^2}\right)}{1000 \, \dfrac{\text{lbf}}{\text{kip}}}$$

$$= 7.7 \text{ kips}$$

The tributary area, A_T, for live loads is

$$A_T = (\text{floors affected})A = (3)(256 \text{ ft}^2) = 768 \text{ ft}^2$$

From ASCE/SEI7 Table 4-2, the live load element factor, K_{LL}, for an exterior column without cantilever slabs is 4.

$$K_{LL}A_T = (4)(768 \text{ ft}^2) = 3072 \text{ ft}^2 \geq 400 \text{ ft}^2$$

From ASCE/SEI7 Sec. 4.7.2, members for which a value of $K_{LL}A_T$ is 400 ft² or greater may have live loads reduced using ASCE/SEI7 Eq. 4.7-1. L_o is the unreduced design live load, and L is the reduced design live load per square foot.

$$L = L_o \left(0.25 + \frac{15}{\sqrt{K_{LL}A_T}} \right)$$

$$= \left(50 \ \frac{\text{lbf}}{\text{ft}^2} \right) \left(0.25 + \frac{15}{\sqrt{(4)(768 \ \text{ft}^2)}} \right)$$

$$= 26 \ \text{lbf/ft}^2 \quad [\text{controls}]$$

L cannot be less than $0.40L_o$ for members supporting two or more floors.

$$0.40L_o = (0.40) \left(50 \ \frac{\text{lbf}}{\text{ft}^2} \right) = 20 \ \frac{\text{lbf}}{\text{ft}^2} \quad [\text{does not control}]$$

The load at the footing due to live load, P_L, is

$$P_L = L A_T = \frac{\left(26 \ \dfrac{\text{lbf}}{\text{ft}^2} \right)(768 \ \text{ft}^2)}{1000 \ \dfrac{\text{lbf}}{\text{kip}}} = 20 \ \text{kips}$$

To find the load at the footing due to horizontal seismic loads, first find the overturning moment, M, at the base of the braced frame. F_n is the horizontal seismic force at floor n, and h_n is the height of floor n.

$$M = F_1 h_1 + F_2(h_1 + h_2)$$
$$+ \cdots + F_n(h_1 + h_2 + \cdots + h_n)$$
$$= (11 \ \text{kips})(15 \ \text{ft}) + (20 \ \text{kips})(15 \ \text{ft} + 12 \ \text{ft})$$
$$+ (30 \ \text{kips})(15 \ \text{ft} + 12 \ \text{ft} + 12 \ \text{ft})$$
$$+ (39 \ \text{kips})(15 \ \text{ft} + 12 \ \text{ft} + 12 \ \text{ft} + 12 \ \text{ft})$$
$$= 3864 \ \text{ft-kips}$$

The axial load at the footing due to horizontal seismic loads, P_{Q_E}, is equal to the overturning moment, M, divided by the column spacing, B.

$$P_{Q_E} = \frac{M}{B} = \frac{3864 \ \text{ft-kips}}{30 \ \text{ft}} = 128.8 \ \text{kips}$$

Therefore, the maximum factored reaction, P_u, at footing B-1 is

$$P_u = (1.2 + 0.2S_{DS})P_D + P_{Q_E} + 0.5P_L + 0.2P_S$$
$$= (1.2 + (0.2)(0.40))(153.6 \ \text{kips})$$
$$+ 128.8 \ \text{kips} + (0.5)(20 \ \text{kips}) + (0.2)(7.7 \ \text{kips})$$
$$= \boxed{337 \ \text{kips}}$$

88. (b) The punching shear strength of a concrete footing, V_c, is determined in accordance with ACI 318 Sec. 22.6.5.2. For a column with roughly square dimensions, ACI 318 Table 22.6.5.2(a) controls.

$$\phi V_c = \phi 4\lambda \sqrt{f_c'} \, b_o d$$

From ACI 318 Table 19.2.4.2, for normal weight concrete,

$$\lambda = 1.0$$

From ACI 318 Table 21.2.1, for shear and torsion,

$$\phi = 0.75$$

Use ACI 318 Sec. 13.2.7.2, Sec. 13.2.7.1, and Sec. 22.6.4.1 to determine the critical section for shear. For a column with a steel baseplate, the critical section for shear is taken at a distance $d/2$ away from halfway between the face of the column and the edge of the steel baseplate. This location for the critical shear perimeter must be used at three sides of the column.

For the side of the column with a gusset, ACI 318 does not explicitly address how to determine the distance to the critical shear perimeter, so judgment must be used. The illustration shown offers one valid assumption for the perimeter of the critical section for b_o. However, other more conservative assumptions for the critical shear perimeter could also be used.

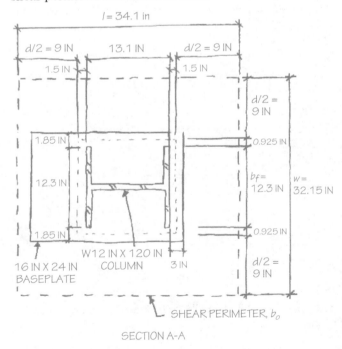

SECTION A-A

From AISC *Steel Construction Manual* Table 1-1, a $W12 \times 120$ column has dimensions of $d = 13.1$ in and $b_f = 12.3$ in. The baseplate is 24 in $\times$ 16 in.

$$
\begin{aligned}
b_o &= 2l + 2w \\
&= (2)(9 \text{ in} + 1.5 \text{ in} + 13.1 \text{ in} + 1.5 \text{ in} + 9 \text{ in}) \\
&\quad + (2)(9 \text{ in} + 0.925 \text{ in} + 12.3 \text{ in} + 0.925 \text{ in} + 9 \text{ in}) \\
&= 132.5 \text{ in}
\end{aligned}
$$

The vertical load on the footing was calculated in problem part 88(a) as $P_u = V_u = 337$ kips and $d = 18$ in per problem illustration III.

The punching shear strength of the concrete footing is

$$
\begin{aligned}
\phi V_c &= \phi 4\lambda\sqrt{f_c'}\, b_o d \\
&= \frac{(0.75)(4)(1.0)\sqrt{4000 \dfrac{\text{lbf}}{\text{in}^2}}\,(132.5 \text{ in})(18 \text{ in})}{1000 \dfrac{\text{lbf}}{\text{kip}}} \\
&= 453 \text{ kips} > 337 \text{ kips} \quad [\text{OK}]
\end{aligned}
$$

> The footing is adequate for punching shear.

(Note that the use of the full footing load of $V_u = 337$ kips is conservative for a punching shear check. Although not required, the portion of the factored pressure, q_u, could be subtracted within the critical shear perimeter, and a reduced value for V_u could be used.)

B is the width of the footing, and L is the length of the footing from problem illustration III.

Find the factored pressure, q_u.

$$
q_u = \frac{P_u}{LB} = \frac{337 \text{ kips}}{(5 \text{ ft})(5 \text{ ft})} = 13.48 \text{ kips/ft}^2
$$

The reduced value for V_u is

$$
\begin{aligned}
V_u &= P_u - q_u lw \\
&= 337 \text{ kips} - \left(13.48 \frac{\text{kips}}{\text{ft}^2}\right)\left(\frac{(34.1 \text{ in})(32.15 \text{ in})}{\left(12 \dfrac{\text{in}}{\text{ft}}\right)^2}\right) \\
&= 234 \text{ kips} < 453 \text{ kips} \quad [\text{OK}]
\end{aligned}
$$

88. (c) Per AISC 341 (Part 9.1 of the AISC *Seismic Construction Manual*) Sec. F2.3, the expected brace strength in tension is

$$
P_t = R_y F_y A_g
$$

From AISC 341 Table A3.1, for ASTM A500 steel, $R_y = 1.4$.

From AISC *Steel Construction Manual* Table 1-12, for an HSS6 $\times$ 6 $\times$ ⅜ in shape, $A_g = 7.58$ in². Therefore, the expected brace strength in tension is

$$
P_t = R_y F_y A_g = (1.4)\left(46 \frac{\text{kips}}{\text{in}^2}\right)(7.58 \text{ in}^2) = 488 \text{ kips}
$$

From AISC 341 Sec. F2.3, the expected brace strength in compression is the smallest of

$$
P_c = \min\begin{cases} R_y F_y A_g \\ 1.14(F_{cre} A_g) \end{cases}
$$

F_{cre} is the critical buckling stress determined from AISC 360 Chap. E using the expected yield stress $R_y F_y$ in lieu of F_y.

From the problem statement, use workpoint-to-workpoint dimensions. (The actual brace length will be less, due to the size of the gusset plates.) For members pinned at their ends, K is 1.0. From AISC *Steel Construction Manual* Table 1-12, for an HSS6 $\times$ 6 $\times$ ⅜ in shape, $r = 2.28$ in. Therefore, the slenderness ratio, KL/r, of the brace is

$$
\begin{aligned}
\frac{KL}{r} &= \frac{(1.0)\sqrt{(15 \text{ ft})^2 + (15 \text{ ft})^2}\left(12 \dfrac{\text{in}}{\text{ft}}\right)}{2.28 \text{ in}} \\
&= 111.6
\end{aligned}
$$

From AISC 360 Sec. E3, replacing F_y with $R_y F_y$,

$$
\begin{aligned}
4.71\sqrt{\frac{E}{R_y F_y}} &= 4.71\sqrt{\frac{29{,}000}{(1.4)\left(46 \dfrac{\text{kips}}{\text{in}^2}\right)}} \\
&= 99.9 < KL/r
\end{aligned}
$$

Since the slenderness ratio is greater than $4.71\sqrt{E/R_y F}$, use AISC 360 Eq. E3-3,

$$
F_{cre} = 0.877 F_e
$$

Per AISC 360 Eq. E3-4,

$$
\begin{aligned}
F_e &= \frac{\pi^2 E}{\left(\dfrac{KL}{r}\right)^2} = \frac{\pi^2\left(29{,}000 \dfrac{\text{kips}}{\text{in}^2}\right)}{(111.6)^2} \\
&= 23.0 \text{ kips/in}^2
\end{aligned}
$$

Per AISC 360 Eq. E3-3,

$$F_{cre} = 0.877F_e = (0.877)\left(23.0 \ \frac{\text{kips}}{\text{in}^2}\right)$$
$$= 20.2 \ \text{kips/in}^2$$

Therefore, the expected brace strength in compression is

$$P_c = \min\begin{cases} R_y F_y A_g = 488 \ \text{kips} \\ 1.14(F_{cre}A_g) = (1.14)\left(20.2 \ \frac{\text{kips}}{\text{in}^2}\right)(7.58 \ \text{in}^2) \\ \qquad\qquad = 175 \ \text{kips} \quad [\text{controls}] \end{cases}$$

$$\boxed{P_t = 488 \ \text{kips}}$$

$$\boxed{P_c = 175 \ \text{kips}}$$

Determine the maximum moment demand on beam 1. From AISC 341 Sec. F2.3, a maximum of $0.3P_c$ is used as the expected post-buckling brace strength.

$$P_{c,\text{buckling}} = 0.3P_c = (0.3)(175 \ \text{kips})$$
$$= \boxed{52.5 \ \text{kips}}$$

Also from AISC 341 Sec. F2.3, the beams in a special concentrically braced frame must be evaluated for the applicable load combination including the amplified seismic load (ASCE/SEI7 Sec. 12.4.2.3, load combination 5). The amplified seismic load is the larger force determined from the following cases.

case 1: an analysis in which all braces are assumed to resist forces corresponding to their expected strength in compression or in tension

case 2: an analysis in which all braces in tension are assumed to resist forces corresponding to their expected strength, and all braces in compression are assumed to resist their expected post-buckling strength

By inspection, the second case will produce a larger unbalanced vertical force at the midspan of beam 1.

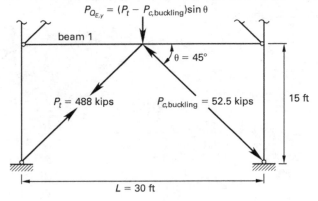

The unbalanced vertical force at the midspan of beam 1 due to the seismic forces is

$$P_{Q_{E,y}} = (P_t - P_{c,\text{buckling}})\sin\theta$$
$$= (488 \ \text{kips} - 52.5 \ \text{kips})\sin 45°$$
$$= 308 \ \text{kips}$$

The braces will not provide support for gravity loads in the post-buckled condition, so consider beam 1 as simply supported with a span of $L = 30$ ft for a moment due to dead and live loads.

The moment at the midspan of beam 1 due to seismic force, M_{Q_E}, is

$$M_{Q_E} = \frac{P_{Q_{E,y}}L}{4} = \frac{(308 \ \text{kips})(30 \ \text{ft})}{4}$$
$$= 2310 \ \text{ft-kips}$$

The moment due to the given distributed dead load, M_D, is

$$M_D = \frac{w_{\text{dead}}L^2}{8} = \frac{\left(750 \ \dfrac{\text{lbf}}{\text{ft}}\right)(30 \ \text{ft})^2}{(8)\left(1000 \ \dfrac{\text{lbf}}{\text{kip}}\right)} = 84.4 \ \text{ft-kips}$$

The moment due to the given distributed live load, M_L, is

$$M_L = \frac{w_{\text{live}}L^2}{8} = \frac{\left(250 \ \dfrac{\text{lbf}}{\text{ft}}\right)(30 \ \text{ft})^2}{(8)\left(1000 \ \dfrac{\text{lbf}}{\text{kip}}\right)} = 28.1 \ \text{ft-kips}$$

Use ASCE/SEI7 Sec. 12.4.2.3, load combination 5. Neglecting the snow load, the factored moment demand, M_u, is

$$M_u = (1.2 + 0.2S_{DS})M_D + M_{Q_{E,y}} + 0.5M_L$$
$$= (1.2 + (0.2)(0.4))(84.4 \ \text{ft-kips}) + 2310 \ \text{ft-kips}$$
$$\qquad + (0.5)(28.1 \ \text{ft-kips})$$
$$= \boxed{2430 \ \text{ft-kips}}$$

88. (d) The large moment demand on beam 1 is primarily due to the unbalanced vertical seismic force of 308 kips from the braces. Three options for reconfiguring the braces in order to reduce or eliminate the unbalanced vertical force are as follows. (Only one configuration and sketch is required for full credit.)

1. Alternate the orientation of the braces (inverted V

and V) at each floor in order to reduce the net vertical seismic force at the midspan of beam 1.

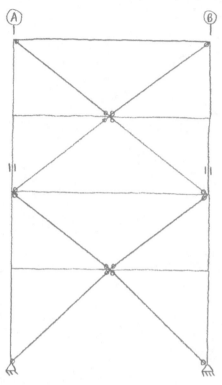

2. Use a cross-braced frame system where the braces extend the full width of the bay.

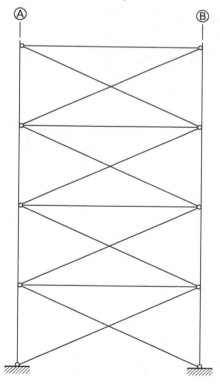

3. Introduce a zipper column that extends from the first floor to the roof.

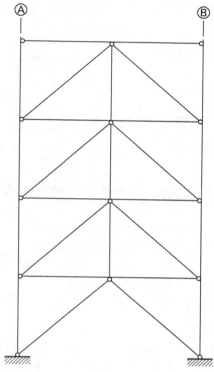

88. (e) The required strength of brace connections for special concentrically braced frames is given in AISC 341 Sec. F2.6c. By inspection, the required tensile strength will control over the required compressive strength for the design of the weld.

From AISC 341 Sec. F2.6c(1), the required tensile strength is the lesser of the expected yield strength of the brace in tension and the maximum load effect that can be transferred to the brace by the system. In the absence of information about the maximum load effect that can be transferred to the brace, design the weld for the expected yield strength.

The expected yield strength of an HSS6 × 6 × ⅜ in brace was calculated in problem part 88(c) as

$$P_t = 488 \text{ kips}$$

Use a weld size of ⁵⁄₁₆ in, as this will meet the minimum fillet weld size requirement of AISC 360 Table J2.4 (regardless of the gusset plate thickness).

D is the weld size in sixteenths-of-an-inch, and l is the weld length in inches. For LRFD, the available strength of an E70XX fillet weld is given in AISC *Steel Construction Manual* Part 8 as

$$\phi R_n = \left(1.392 \; \frac{\text{kips}}{\text{in}}\right) Dl$$

For a brace-to-gusset slotted connection, there will be four equal-length welds. Setting the yield strength equal to the weld strength gives a required weld length, l, of

$$l = \frac{P_t}{(\text{no. of welds})\phi R_n D} = \frac{488 \text{ kips}}{(4)\left(1.392 \dfrac{\text{kips}}{\text{in}}\right)(5)}$$

$$= 17.5 \text{ in}$$

Use four 18 in long, $\frac{5}{16}$ in fillet welds at the brace-to-gusset connection.

(There are alternative acceptable designs, such as using four 15 in long, $\frac{3}{8}$ in fillet welds.)

88. (f) Powder-actuated fasteners may not be installed in the gusset plates. From AISC 341 Sec. F2.5c., the protected zone of the special concentrically braced frame includes elements that connect braces to beams and columns. From AISC 341 Sec. I2.1, "welded, bolted, screwed, or shot-in attachments (e.g., powder-actuated fasteners) for ... duct work ... shall not be placed within protected zones."

Refer to AISC 341 Comm. Fig. C-F2.5 for an illustration of the protected zone of an inverted V-braced frame.